MIMO Antennas for Wireless Communication

MIMO Antennas for Wireless Communication

Theory and Design

Leeladhar Malviya

Rajib Kumar Panigrahi

M.V. Kartikeyan

CRC Press
Taylor & Francis Group
Boca Raton London New York

CRC Press is an imprint of the
Taylor & Francis Group, an **informa** business

First edition published 2021
by CRC Press
6000 Broken Sound Parkway NW, Suite 300, Boca Raton, FL 33487-2742

and by CRC Press
2 Park Square, Milton Park, Abingdon, Oxon, OX14 4RN

First edition published by CRC Press 2021

CRC Press is an imprint of Taylor & Francis Group, LLC

ISBN: 9780367530471 (hbk)
ISBN: 9780367530532 (pbk)
ISBN: 9781003080275 (ebk)

Typeset in Times
by KnowledgeWorks Global Ltd.

*To our Parents
and Teachers.*

Contents

Preface

We are delighted to introduce the book on MIMO Antennas for Wireless Communication: Theory and Design. Multiple-input-multiple-output (MIMO) antennas are the current hot topic of research and technology. Next generations of wireless antenna designs are very keen to utilize the aspects of MIMO in a variety of ways to accommodate all generations of wireless and mobile needs with compact and portable devices. Most of the literature is available on the internet, research papers, and in books about the physical layer product of internet protocol and its specification dispersively. We are pleased to introduce all of them inside the book in a manner that strengthens the understanding of beginners and designers of MIMO antennas.

Wireless communication with the antennas at the physical layer of internet protocol is the direct source of data communication between the base station and mobile station. The separation between the transmitter and receiver leads to the multipath propagation. Due to multipath propagation, the problem of line-of-sight communication can be solved with the help of MIMO antennas. These are essential part of wireless communication and now the heart of wireless technology. Wireless communication is nothing without it. MIMO is a direct outgrowth of data rate, bandwidth saving, and capacity enhancement.

MIMO was developed with the vision to strengthen the link reliability, user specific data rate, bandwidth on demand, video on demand, and various other services in rich multipath environments. Enhancement in spectral efficiency with the generation of MIMO antenna technology opened the ways to save the bandwidth and covered all the hidden parts of user specific areas. Also, the massive MIMO antenna usage in 5G communication aims to strengthen the services for wireless technology users.

MIMO antennas on the printed circuit board are now frequently available with variety of shapes and orientations to provide required coverage with a very low latency, and their costs are in reach of users. In MIMO research fields, this book can play a significant role in bringing some important aspects to the antenna design field.

Twenty-first century wireless communication is witnessing a miraculous evolution in technology. There has been a tremendous advancement in technology the first generation of analog communication and larger-size phones/devices to digital communication with compact-size phones/devices and has influenced users of every class.

Immense data rate, large capacity, ample coverage, large bandwidth, low power consumption, excellent quality of services, link reliability, low latency, voice/video/bandwidth on demand etc. are the contemporary technology demands essential for high-definition televisions, multimedia applications, mobile phone applications, and in internet services. Modern and smart technology demands Gbps data rates for indoor and outdoor applications and uninterrupted communication in line-of-sight as well as in non-line-of-sight communications due to the mobility of users. Fulfillment of such requirements accompanying a single input transmitter and single output receiver (SISO) calls for very large bandwidth and excessive power consumption. Therefore, the SISO system is not suitable for such requirements.

MIMO addresses all the above problems and is an advanced technology for wireless local area network (WLAN), WiMAX, long-term evolution (LTE), 5G, and future generations. The present and forthcoming generations of wireless communication systems require all the capabilities to incorporate voice, video, image, and data services within the portable and mobile devices. MIMO has such capabilities and can be integrated within compact devices for indoor and outdoor activities. The problem of signal fading and coverage in rich multipath environments are taken care by employing multiple transmit/receive antennas. Massive MIMO communication is one of the key technologies to enhance the performance of wireless terminals to cover the existing needs.

Present/future wireless technology support portable and miniaturized devices. MIMO antennas with the limited space have absolute drawbacks. The integration of multiple antennas closely in a small and compact device with the maintenance of very high isolation or negligible mutual coupling is rather complicated with the shared ground and conducting parts. To maximize the performance of MIMO channel link, each transmitter and receiver must be statistically independent.

The integration of MIMO with diversity techniques enhances the other capabilities like coverage, isolation, diversity gain, MIMO gain, gain, effciency, and counts up assured percentage in miniaturization. Diversity effects may be constructive on some specific locations but not on all the places. All the features are unable to be satisfied with a specific diversity technique. Therefore, different diversity techniques are practiced.

Based on the organized literature review, we discover that variety of MIMO antennas are available; however, achieving the low correlation along with compact size, and low envelop correction coefficient (ECC), in a single MIMO antenna design is still a challenge. Also, the coverage of most of the wireless bands is mostly limited to 2×1 MIMO antennas. Few MIMO antennas with circular polarization (CP) have been reported in the literature. Simultaneously, achieving a compact size with low correlation and ECC in circularly polarized MIMO antennas is still a challenge. Therefore, compact MIMO antennas with better performance characteristics can be designed for present and future wireless applications based on distinct radiation patterns, based on unification of vertical polarization and horizontal polarization, and with CP.

The design based on polarization diversity has been presented in Chapter 7. It consists of 2 × 2 MIMO antenna with polarization diversity and with partially stepped ground for 2.4/2.5/5.2/5.5 GHz WLAN/WiMAX standards under the 2.37–2.69 GHz and 4.89–5.61 GHz frequency ranges. The design achieves a very low value of ECC for VSWR ≤ 2 in both the frequency bands.

The coverage of the intended domain requires special attention, because of continuous signaling in live services. For such applications, pattern diversity is an adequate technique among existing diversity techniques. The pattern diversity technique achieves complementary field patterns for properly implemented cases. A 2 × 2 MIMO antenna accompanying pattern diversity and partially extended ground modification technique is designed for wide-band operation to accommodate 2.4/2.5/3.5/5.2/5.5/5.8 GHz WLAN/WiMAX applications under the frequency range of 2.0–7.31 GHz in Chapter 8. The design fulfills all the set goals for the intended application.

In many cases receiver is in a confusing state due to the polarization of transmitting antennas and also when one/both of the communicating antennas are mobile. In such cases, CP is accepted over the linear polarization (LP). In CP MIMO antennas, every radiator is adequate in receiving signals, having any polarization of transmitter. That in turn resolves the problem of multipath propagation in all weather conditions regardless of the type of fading and diversity techniques. CP antennas radiate and penetrate in all planes in vertical as well as in horizontal directions. CP polarization is achieved with two asymmetric slots etched at the center of a considered patch in Chapter 9. A 2 × 2 MIMO antenna with two 1 × 2 feed arms is designed for CP without any mutual coupling reduction technique for IEEE 802.11 WLAN.

The 4G LTE provides increased data rates, higher signal-to-noise ratios (SNRs), and low latency using MIMO antenna technology. Number of operating bands of LTE are decided to cover uplink and downlink frequencies with scalable bandwidths and having backward compatibility with WLAN/WiMAX and many other wireless standards. Jointly, LTE and MIMO fulfill the voice/video/multimedia on-demand services and provide high SNRs. Two 2 × 2 MIMO antennas with diversity effects are designed for LTE 1800/1900 applications to cover 1.6–2.2 GHz frequency band in Chapter 10.

Chapter 11 presents two MIMO antenna designs for wireless applications. In first design, pentagon-shaped patches with T-shaped isolator and diversity effects is presented for 2.4/2.5/3.5 GHz WLAN/WiMAX applications. In second design, the compact MIMO antenna design with diversity effects is presented with power divider arms. Effect of the ground stub results in lower mutual coupling between ports. Effect of the equivalent circuit shows the tunability of the design. Designed MIMO is used for 2.23–2.64 GHz and 3.26–3.70 GHz frequency bands, and is applicable for both the indoor and outdoor environments.

The desired objective of the present research is concerned with the investigation of diversity and mutual coupling effects on presented MIMO antenna designs and their study. This book demonstrates the accomplishment of diversity techniques in MIMO antennas that leads to the improvement of performance ratings such as return-loss, isolation, gain, efficiency, ECC, total active reflection coefficient, size of the radiating element, and impedance matching etc.

Comprehensively, the book contributes toward the development of MIMO antennas with diversity techniques for high data rate, better quality of services, and high SNR. The improved MIMO antenna structures are investigated and presented in this book.

During the course of the preparation of this book, our colleagues at Indian Institute of Technology Roorkee helped us immensely. We sincerely thank Prof. Dharmendra Singh, Prof. A. Patnaik, Prof. M. P. S. Chawla, Prof. P. D. Vyavahare, Prof. Vineeta Choudhary, Late Anurag Shrivastava, Dr. Pravin Prajapati, Dr. Arjun Kumar, Dr. Jagannath Malik, Dr. Gaurav Singh Baghel, Dr. Sukwinder Singh, Dr. Himanshu Maurya, Dr. S. Yuvaraj, Dr. Ajeet Kumar, Dr. Kumar Goodwill, Dr. Ramesh Patel, Dr. S. Pelluri, Ms. Surbhi Adya, Mr. Debasish Mondal, Mr. Aditya Singh Thakur, and Mr. Avinash. Thanks are also due to Ms. I. Suryarajitha for preparing the front-page artwork meticulously. Special thanks are due to Shri G. S. Institute of Technology and Science, Indore (M.P.), Indian Institute of Technology Roorkee (IIT Roorkee), and Quality Improvement Programme (QIP)-AICTE Government of India for their generous support and encouragement.

Leeladhar Malviya

R. K. Panigrahi

M. V. Kartikeyan

Acknowledgments

We sincerely thank the authorities of IIT-Roorkee for their kind permission to come up with this book. In addition, special thanks are due to all the authors of the original sources for letting us use their work in this book.

We sincerely thank the following journals/publications/publishers for their kind permission and for the use of their works and reprint permission (permission details are given in the respective references and the corresponding citations are duly referred in the captions):

–Electromagnetics Academy.
–John Wiley & Sons Inc.
–EuMA/Cambridge University Press.
–Taylor and Francis Inc.
–International Journal of Microwave and Optical Technology.

Symbols

Symbol Description

θ	Azimuthal angle
C	Channel capacity
H	Channel matrix
h	Channel parameter
η	Efficiency
ϕ	Elevation angle
ρ_e	Envelope correlation coefficient
G_θ	Gain with respect to azimuthal angle θ
G_ϕ	Gain with respect to elevation angle ϕ
S_{ij}/S_{ji}	Isolation or Mutual coupling between i^{th} and j^{th} ports and vice versa
$tan\delta$	Loss tangent
N_r	Number of receivers
N_t	Number of transmitters
Ω	Ohm (Unit of resistance)
p_θ	PDF with respect to azimuthal angle θ
p_ϕ	PDF with respect to elevation angle ϕ
μ	Permeability
ϵ	Permittivity
θ_1^o	Phase difference between major and minor slots
ϵ_r	Relative dielectric constant
S_{ii}/S_{jj}	Return-loss at i^{th} or j^{th} port
f_r	Resonant frequency
Ω	Solid angle
Γ_a^t	Total active reflection coefficient
λ	Wavelength
σ	Variance

1

Introduction

In the modern scenario, wireless communication is the most demanding and fastest extending application. Last decade witnessed the exponential growth of wireless applications like multimedia, voice over internet protocol (VoIP), bandwidth/video on demand (BOD/VOD), automatic traffic signaling in industry, research, smart home appliances, and personnel services. Hence, researchers and industrialists are attracted to produce portable and compact radio frequency (RF) devices on large scale. Certain applications like mobile phone and bluetooth have become the essential need of wireless users. The physical layer of internet protocol (IP) i.e. IPv4, IPv6 etc., uses the RF antennas to accomplish duplex communication for continuous, reliable, and dedicated links. Multiple-input multiple-output (MIMO) antenna technology has evolved as an essential part of fourth and fifth generations of wireless applications due to its advantages over a single radiating element on low cost printed circuit boards (PCBs). Involvement of MIMO antennas leads to the high data rate, low power consumption, spectrum saving, large capacity, and better quality of services (QOS) in non-line-of-site (NLOS) communication. MIMO antennas have a lot of capabilities and qualities but are highly affected by the correlation of closely packed radiators, which is one of the challenges in MIMO antenna design.

1.1 Fundamentals of MIMO Antennas

Modern wireless and mobile applications with low cost and lightweight antennas are integrated within the microwave circuits. MIMO consists of microstrip patch antennas (MPAs) and is implemented on a PCB with a thin layer of perfect electric conductor (PEC) over the dual sides of the dielectric substrate like FR-4 and Rogers RT etc.

MPA is a physical layer radiating product and an essential part of duplex communication [1,2]. MPAs have been utilized in both civil and military applications due to low cost, light weight, low volume, multi-banding, and easy integrability with microwave components [3–11]. On the other hand, MPAs have limited bandwidth, low power handling, polarization disparities, and low gain capabilities [12]. Thick substrates may produce large bandwidth and high efficiency, but at the cost of power loss due to surface wave generation [13].

MIMO antenna technology offers required data rates for base station (BS), automobile, pedestrian, satellite, and radar applications. Modern and smart technologies with MIMO antenna systems are reported for wireless local area network (WLAN) licensed/unlicensed bands, wireless inter-operability for microwave access (WiMAX) bands, global system for mobile communication (GSM), long-term evolution (LTE), global positioning system (GPS) applications, 5G, and future generations, etc. [14,15].

MIMO system came in light due to limitations of single input single output (SISO) system. The Gigabit per second (Gbps) data rate with SISO requires very large-frequency spectrum and is even unable to cooperate in NLOS networks. Also, SISO requires a very large signal to noise ratio (SNR) in practical receivers. Achieving average signal to interference with noise ratio (SINR) $>$ 10 dB is very critical in SISO, even with spectral efficiencies $\geq$4 bits/s/Hz. Also, due to the power constraint, high gain SISO results in scattering and is unable to assist in NLOS. Extending bandwidth in SISO to achieve Gbps data rate limits the range and reduces fade margin. Hence, large power is required for large coverage. SISO has a range reduction of three times in comparison with modern wireless devices. Frequency reuse plan using SISO requires greater than five times link bandwidth in modern wireless systems [16–19].

Wireless and mobile applications have witnessed considerable progress from analog communication (first generation) to the digital communication (4th generation). Main challenges with modern antenna designs are space, interoperability, multi-banding, specific absorption rate (SAR), and hearing aid capability (HAC). Variable data rate, high capacity, and scalable bandwidth are few other essential requirements in digital communications at the base and mobile stations. MIMO antenna behavior is characterized by far-field gain, diversity gain (DG), envelope correlation coefficient (ECC), total active reflection coefficient (TARC), mean effective gain (MEG), effective diversity gain (EDG), capacity, and antenna efficiency [20]. MIMO analysis and modeling are essential, irrespective of the type of channel [21–27].

Modern MIMO designs are concerned with wireless setups for indoor/outdoor activities with different mediums. Hence, site planning, polarization, number of radiators, size/volume of PCBs, angle diversity, compactness, current coupling, and their control must be investigated very carefully [28–31]. Modern research is better concerned with MIMO antennas in multipath propagation with improved features.

MIMO with limited space and large number of radiators takes care of performance over the wireless environment. There is no space constraint at the base station to accommodate large number of radiators due to multiple wavelength (λ) separation between antennas. Whereas, $\leq 0.5\lambda$ is available in mobile and portable devices. Sufficient space is available even for laptops, personal digital assistants (PDAs), and some other portable devices to make them uncorrelated. Diversity and space reduction techniques limit the space requirement and improve the performance of MIMO applications [32].

Various diversity techniques take care of signal continuity, link reliability, QOS, and coverage in a particular direction etc. Space, pattern, and polarization diversity techniques are generally used in practice. Space diversity is very easy and completely depends on space variation between radiators. Space diversity can easily be implemented at the base stations due to large space available between the radiators. Due to limited space, space diversity is not beneficial at the mobile station and demands large size of implementation. Polarization diversity on the other hand takes care of signals in vertical and horizontal directions. Placement of radiators in orthogonal position ensures required isolation and reduces the overall size of the implementation. Similarly, pattern diversity takes care of unending signaling in desired directions by creating various far-field radiation patterns. It contributes complementary field patterns for properly implemented MIMO designs.

Linearly polarized antennas are easily affected by polarization changes. In cases where polarization is not sure, circularly polarized antennas are required to limit polarization losses and maintain the link reliability. Various diversity techniques can be combined with de-correlation techniques to improve MIMO design parameters.

1.2 Motivation and Scope

Huge demand for MIMO antennas has attracted researchers, academicians, and industrialists. Various wireless technologies have been integrated with MIMO technology to enhance data rates, capacity, and to limit bandwidth and power in NLOS communication. MIMO satisfies the modern need of variable data rates and scalable bandwidth for each wireless application. However, due to limited space, the issue of mutual coupling requires special attention. Diversity and mutual coupling reduction techniques are therefore required.

There is not much complexity and requirements for 2×1 radiating elements on PCB. Only two elements are coupled in this case. However, 2×2 MIMO designs have four radiators and a number of mutual coupling paths. Therefore, strong MIMO designs are required to control the coupling and to enhance design parameters.

Therefore, wireless and mobile engineers are trying their best for efficient and advanced MIMO antenna designs like massive MIMO to satisfy variable Gbps data rates and large capacity. This fact motivated to take up these problems for the enhancement of MIMO antenna performance and to take up this research of using diversity techniques for achieving the same.

The main idea of our book is to bring awareness to researchers in the field of MIMO and massive MIMO antennas. The following points are observed in a thorough investigation of MIMO antenna designs:

1. A number of MIMO designs are available; however, achieving high isolation, compact size, high gain, low ECC, effective active bandwidth, and low MEG in a single MIMO antenna design is still a challenge.

2. Coverage of most of the WLAN/WiMAX bands is mostly limited to 2×1, due to large complexity with 2×2 MIMO antennas available for wireless communication. Increase in the number of data streams makes the analysis and processing very complex.

3. Very few MIMO antennas with circular polarization have been reported in the literature. Simultaneously, achieving compactness with high isolation in circularly polarized MIMO antennas is still a challenge.

4. Always using mutual coupling reduction techniques is not the best solution. Therefore, diversity effects have been utilized and such shapes are generated, which leads to desired outcomes in terms of mutual coupling and size.

The following investigations are undertaken for the performance enhancement of MIMO antennas with different diversity techniques:

1. To elaborate 2×2 compact MIMO antenna design methodology with diversity technique for low mutual coupling for WLAN/WiMAX wireless communication.

2. To explore and develop the design methodology of 2×2 compact MIMO antenna for wide bandwidth with link reliability and low mutual coupling for wireless systems.

3. To develop the design approach of 2×2 MIMO antenna with circular polarization technique with low mutual coupling for WLAN application.

4. To develop the design method of compact 2×2 MIMO with low mutual coupling for long-term evolution applications.

5. To design and develop MIMO antennas with power divider arms and diversity effects for wireless applications.

1.3 Organization of the Book

This book is divided into twelve chapters including the present chapter and is organized as follows:

In **Chapter 1**, fundamentals of MIMO, the motivation of research, research objectives, problem statement, and the organization of book are discussed in detail.

In **Chapter 2**, the theory of MIMO, wireless channel limitations with fading and interference, and approaches of capacity enhancement are discussed.

In **Chapter 3**, functions of MIMO, types of MIMO, and applications of MIMO are covered.

In **Chapter 4**, MIMO antenna performance parameters like reflection coefficient, VSWR, transmitted and reflected powers, transmission coefficient, isolation, ECC, TARC, and MEG are discussed.

In **Chapter 5** massive MIMO antenna system with various frequency bands in 5G, channel estimation, overheads in TDD/FDD, spatial diversity/multiplexing techniques, and different types of beamforming techniques are discussed.

Chapter 6 reviews various diversity and mutual coupling reduction techniques, and various miniaturization techniques are also discussed here thoroughly.

Chapter 7 describes the introduction and related work of multi-band MIMO antenna design with G-shaped folded radiating structure, implementations of it with diversity and partially stepped ground, and different performance parameters including indoor and outdoor criteria.

Chapter 8 describes the introduction and related work of wide-band MIMO antenna, implementations of it with diversity and partially extended ground, and different performance parameters including indoor and outdoor conditions.

Chapter 9 describes the introduction and related work of CP-MIMO antenna, implementations of it with LHCP and RHCP, and various design parameters with indoor and outdoor environmental conditions.

Chapter 10 describes two MIMO antenna designs. First design consists of a folded loop-shaped radiating structure, and the second design uses mathematically inspired patch shape for MIMO implementation. Introduction and related works, implementations of LTE MIMO antenna designs, and experimental results of each design are discussed here with indoor and outdoor criteria.

Chapter 11 describes two MIMO antenna designs with power divider arms for wireless applications. First design consists of pentagon-shaped radiator with inverted L-shaped slot, and the second design has very compact radiator with inverted L-shaped slot for MIMO implementations. Introduction and related works, implementations of MIMO antenna designs, and experimental results of each design are discussed here.

Finally, **Chapter 12** provides the book contributions towards the MIMO antenna designs for modern wireless applications, and the future perspective of the work done.

2

Theory of MIMO

Actual performance of wireless communication devices depends on the distance between the transmitter and receiver. Large-scale and small-scale fadings depend on the propagation of the signal. In general, the signal follows the multipath propagation. Large separations between transmitters and receivers result in large-scale fading, while shorter separations lead to small-scale fading. Effect of co-channel interference and adjacent channel increases the S/I in the system and degrades the actual performance parameters of MIMO antenna. Various channel planning and allocation approaches are used in wireless communication for coverage of each location with antenna placements. By controlling the interference, capacity of the system can be enhanced. The capacity of MIMO antennas depends on the number of streams multiplied by the capacity of single-input single-output (SISO) system.

2.1 Introduction

Electromagnetic (EM) waves between transmitter and receiver pass through multiple paths and severely obstructed by high rise buildings, hills, metal objects etc. The randomness of EM waves makes the channel unpredictable and analysis becomes complex. Two types of fading, i.e. large scale and small scale, are used in practice. Large-scale fading is analyzed by reflection, diffraction, and scattering mechanisms. Small-scale fading includes the fast variations in signal strength over a short distance or time. This is due to the arrival of different signals at the receiver at different times and are caused by EM interference. Due to multipath propagation, resultant EM wave observes variations in signal strength and angle. The capacity of wireless system depends on signal strength (S) to interference (I) ratio. The S/I limits the coverage area by limiting the frequency reuse plan. Finally, the effects of co-channel interference and adjacent channel interference are controlled by improving S. Capacity of wireless channels can be improved by cell splitting, sectoring, repeaters, and microcells. MIMO is the solution for multipath propagation. This chapter describes different fading approaches and capacity improvement methods in detail [33].

2.2 Wireless Channel Limitations

Multipath components severely affect the propagation of waves and limit the actual performance of MIMO antennas. This section discusses large-scale fading, small-scale fading, different types of wireless interference mechanisms, and concept of capacity improvement.

2.2.1 Fading

Large-scale and small-scale fadings and their effects on wireless communication are considered in this section.

2.2.1.1 Large-Scale Fading

Line-of-sight communication in urban areas is not possible all the times due to high rise buildings, mountains, metallic objects, etc. Because of these obstacles, the signal propagates through multiple paths due to reflection, diffraction, and scattering mechanisms. These EM waves interact in medium and produce multipath fading. The signal strength may decrease, as the separation between transmitter and receiver is multiple of several wavelengths (λ). Large-scale fading models predict the signal strength in such large transmitter–receiver separations to predict the channel and coverage area [33, 34].

Propagation of EM waves in the wireless communication channel is highly affected by reflection, diffraction, and scattering mechanisms. On the basis of these mechanisms, received signal strength is predicted. Reflection occurs from high rise buildings, walls, and earth surface. EM waves between transmitter and receiver become a secondary source of waves due to sharp edges of the obstacle surfaces. These secondary waves present in space and behind the obstacles are generated because of bending. At the point of diffraction, these high-frequency waves depend on the dimension of the obstacle, signal strength, angle, and polarity. Scattering occurs when the dimension of the obstacle is very smaller than the λ and obstacles per unit volume are very large. Lamp posts, street signs, and other small objects or irregular channels are the practical examples of scattering [33, 34].

2.2.1.2 Small-Scale Fading

When there are very short distances or time between transmitter and receiver in wireless propagation models, very fast fluctuations in received signal amplitude occur which is known as small-scale fading. Arrival of the transmitted signal at different time intervals from multiple paths at the receiver causes interference (or fading). Resultant EM waves have variation in amplitude, angle, and polarization disparities. Hence, fast changes in signal strength, modulation, and time dispersion create small-scale multipath propagation model.

The height of mobile antennas is much less than the surrounding objects present in urban areas. Therefore, the possibility of line-of-sight communication with the base station is impossible. Also, in line-of-sight, multipath propagation is possible due to reflections of the signal. These randomly delayed signals are combined vectorially and cause fading at the receiver i.e. at mobile. When mobile is stationary and the urban object is in motion, it may cause fading. In the case of static urban models, the motion of mobile is considered only.

Multiple combinations of EM waves cause constructive and destructive phenomena in space. Due to deep fades, mobile may stop at a particular site. This may discontinue the link or dedicated virtual links are broken at this stage. Due to relative motion between transmitter (base station) and receiver (mobile), each arrived signal experiences frequency shift that is called the Doppler shift [33–36].

Following factors influence the small-scale fading:

1. *Multipath propagation:* Multipath reflectors and scatterers in propagation channel build a constantly varying environment to reduce/distort the properties of received signals. Random variations in amplitude and phase of the received signal may cause inter-symbol interference (ISI).

2. *Mobile speed:* Relative motions between transmitting and receiving devices cause randomness in frequency modulation. Hence, the movement of mobile toward or away from the base station decides a positive/negative Doppler shift.

3. *Speed of urban object:* Time-varying Doppler shift is observed if RF channel objects are not stationary. If the speed of urban object is higher than the mobile speed, then small-scale fading occurs.

4. Distortion in received signal may be observed if bandwidth of propagation channel is lower than the transmitted signal bandwidth and vice versa.

2.2.2 Effect of Interference on Channel Capacity

Interference limits the actual performance of the wireless and mobile systems. Presence of another mobile in the same cell, progressing call in neighboring cell, same operating frequency bands of base stations, and signal linkage of another system into the mobile band are the major causes of interference in propagation channel that affect system capacity.

On voice channels, background signal (unwanted transmission of signals) causes cross talk. Whereas, on control channels, errors in signaling cause dropping and blocking of calls. In urban environments, interference becomes very critical due to HF noise and a large number of transmitters (base station) and receivers (mobiles). Such interference limits the maximum value of capacity and call drops are obvious. Due to random propagation of signals, interferences are also generated in the mobile system, and are uncontrollable. Neighborhood of competing base stations causes out-of-band (OOB). Major interferences in mobile systems are recognized by co-channel and adjacent channel interferences [34].

2.2.2.1 Co-Channel Interference and Capacity

Several cells in a given coverage area use the set of same frequencies due to frequency reuse plan. These cells are known as co-channel cells and corresponding signal interference in these cells is called co-channel interference. Increasing the signal power of transmitter does not solve the problem of co-channel interference because increase in signal-to-noise ratio (SNR) at transmitter also increases the interference in co-channel cells. Sufficient isolation between co-channel cells may limit co-channel interference. When all the base stations use equal powers and cells have equal dimensions, then co-channel interference ratio depends on cell radius (R) and separation between closest co-channel cells (D). Hence, ratio Q is utilized to enhance transmission quality and capacity and is given by (2.1) [34].

$$Q = \frac{D}{R} = \sqrt{3N}, \tag{2.1}$$

where Q is the co-channel reuse ratio, and N is the cell cluster size.

Higher the value of Q means lower the co-channel interference and better is the quality of the signal transmission. Also, smaller values of cluster size N lead to larger capacity. Similarly, signal-to-interference ratio S/I may be obtained by (2.2) [34].

$$\frac{S}{I} = \frac{S}{\sum_{i=1}^{i_0} I_i}, \tag{2.2}$$

where i_0 is the number of co-channel interfering cells, S is desired signal power from desired base station, and I_i is the interference power by i^{th} interfering co-channel cell base station. Similarly, average received power is given by (2.3) [34].

$$P_r = P_0 \left(\frac{d}{d_0} \right)^{-n}, \tag{2.3}$$

where P_0 is the received power at small distance d_0 from transmitting antenna, and n is loss exponent.

2.2.2.2 Adjacent Channel Interference and Capacity

If adjacent channel cell is transmitting frequencies, and at the receiver, imperfect filters permit them, then this is known as adjacent channel interference. This effect also referred to the near field, where very close transmitter receives the signal of receiver. Careful channel strategies and powerful filters can limit this effect. Adjacent channel interference may be limited by keeping frequency separation between channels in a particular cell. For large-frequency reuse factors, sufficient separation between the adjacent channels is maintained to limit adjacent channel interference. Practically, high Q cavity filters are required to control adjacent channel interference at base stations [34].

2.2.2.3 Power and Interference

Practically, low power is transmitted by each mobile user to enhance battery life. The transmitting power of mobile is under control of the corresponding base station. For good quality of link, smallest power transmission is necessary.

2.3 Approaches to Improve Capacity

To satisfy the demand of increasing number of users in a cell, more channels per coverage area are required. Hence, capacity of the system can be extended. Certain approaches like cell splitting, sector forming, repeaters, and microcells are used for capacity enhancement. These are described in details here.

2.3.1 Cell Splitting

It solves the problem of the congested cell by dividing it into smaller cells with its base station. It also reduces the antenna height and transmitter power. Enhancement in capacity is observed here due to reutilization of the same channel multiple times. Cell splitting maintains the quality of the transmission link by keeping co-channel reuse factor $Q = D$ at minimum. By decreasing cell radius R, capacity can be enhanced. In practice, finding the real estate is so difficult. Hence, cell splitting is done at the required location only. For cell splitting, the minimum separation between co-channel cells must be maintained for proper channel assignment and interference control. Similarly, by using an umbrella concept, low and high speed users are completely accommodated. This process is also known as the rescaling [34–36].

2.3.2 Sector Forming

Without decreasing the cell radius R, capacity can be increased by frequency reuse and reducing cells in a cluster. But interference must be controlled, without placing a lower limit on transmit power. The co-channel interference may be limited by using a number of directional antennas in place of single omni-directional antenna at the base station. Hence, each directional antenna radiates in a sector, this approach is called sectoring. To cover 360° dimension, each cell is divided into three 120° sectors or six 60° sectors. Out of these sectors, only few sectors receive interference. Hence, interference is limited in this approach. Further improvement in S/I is possible by titling sector antenna return-loss notch in the co-channel cell area. Therefore, it improves the system capacity. This approach results in dedicated channels due to more antennas at base stations. The key difference between sector forming and MIMO is: 3G/UMTS with 3 sectors and 20 element arrays require $3 \times 20 = 60$ antennas, while 4G/LTE-A with 8 MIMO and 30 element arrays require $8 \times 30 = 240$ antennas [34–36].

2.3.3 Repeaters

To extend the range, bi-directional re-transmitters known as repeaters are required. Repeaters receive the signal from the base station, amplify it, and re-radiate it in the region. In this process noise and interference are also amplified and re-radiated. Hence, proper amplification level is maintained. Both the ends of repeaters have directional or distributed antennas for spot coverage in critical locations. Repeaters are the re-radiators only to cover specific location. They do not contribute to capacity of systems [34–36].

2.3.4 Microcell Zones

During handoff in sectoring approach, link load drastically increases. Therefore, microcell zone concept has been adopted. Multiple zones are connected to a single base station via coaxial/fiber-optic cable or microwave link to form a cell. In this approach, antennas are placed at the edges to provide strongest signal to different users. Traveling user with the same channel is retained in each zone here. During this, base station only transfers the channel to the next zone. Due to localization of base station radiation, co-channel interference is controlled. Hence, capacity is improved. This approach is utilized in highways and traffic corridors [34–36].

2.4 Concluding Remarks

In this chapter, different types of fading mechanisms have been given to provide the users with large-scale and small-scale fading concepts. Large-scale fading is analyzed over the large separations between transmitters and receivers, while for shorter separations small-scale fading is used. Different types of interference approaches have also been discussed here. The effect of co-channel interference and adjacent channel interference can be controlled by enhancing S/I. Enhancement in system capacity with channel planning and allocation, with coverage of each location with antenna placements have also been described in detail here. Hence, by controlling the interference, the capacity can be enhanced. The capacity of MIMO is the number of streams multiplied by the capacity of SISO.

3

Applications of MIMO

MIMO is the future of all the wireless generations. MIMO performs three main functions to deal transmission and reception of signal through multipath channels. MIMO precoding, spatial multiplexing, and diversity coding along with their usability in wireless communication are very important to solve the problems related to data rate and coverage of every class of users. Similarly, drawbacks of single-user (SU)-MIMO are solved with multi-user (MU)-MIMO. Massive MIMO is the part of the MU-MIMO. Various applications of MIMO are covered in this chapter.

3.1 Introduction

MIMO is the backbone of the present antenna technology, which provides solution for data rate, capacity, and most of the wireless needs. MIMO functions i.e. precoding, spatial multiplexing, and diversity coding are used to provide variety of solutions. Precoding is used at the transmitter and works on the beamforming for gain maximization. Spatial multiplexing works on channel capacity enhancement strategies. Diversity coding enhances signal diversity. MIMO types include the SU MIMO, MU MIMO, and massive MIMO.

3.2 Functions of MIMO

MIMO functions are divided into following major categories:

1. *Precoding:* MIMO with multiple antennas transmit the same signal at the transmitter to maximize signal power at the receiver. Appropriate values of gain and phase are required in this process. This is called the beamforming or single stream. Precoding is a combination of multiple streams. Hence, is called the multi-stream beamforming. To get the benefit of precoding, transmitter and receiver must have knowledge of channel state information (CSI). Different streams at the receiver add up constructively to enhance receiver gain by controlling the adverse effects of multipath

propagation. Transmit beamforming uses conventional beams and is unable to equalize the gain at the receiver. Therefore, precoding is preferred in modern wireless systems.

2. *Spatial multiplexing (SM):* Wireless communication with the transmission of uniquely encoded and independent data streams using independent antennas is called the spatial multiplexing. At the receiver, these multiple streams are treated as parallel channels due to the spatial signature of each transmitter. Number of streams depends on the number of transmitters. Spatial multiplexing without the CSI enhances channel capacity. Spatial multiplexing with precoding provides maximization of gain at the receiver and capacity enhancement. It can be used in space division multiple access and MU MIMO systems with a variety of transmissions. Maximum spatial multiplexing order is given by (3.1). Parameter N_s increases the spectral efficiency [37].

$$N_s = \min(N_t, N_r) \tag{3.1}$$

where N_t is the number of transmitting antennas and N_r is the number of receiving antennas.

3. *Diversity coding:* In the diversity method, space-time coding is used to encode the signal stream. A single stream is emitted from each transmitter. Diversity coding with multiplexing can be combined to enhance wireless channel performance.

3.3 Types of MIMO

Different types of MIMO are explained in this section.

(i) *Single-user MIMO (SU-MIMO):* Single MIMO transmitter (base station) when communicates with the single MIMO receiver, then it is called a SU-MIMO. SU-MIMO started with the third generation in the year 2009, with 384 kbps data rate. SU-MIMO without CSI is able to provide a high data rate, interference reduction, and high throughput for low signal-to-noise ratio (SNR) conditions. SU-MIMO is suitable for highly complex mobile phones with more receiving antennas [37].

(ii) *Multi-user MIMO (MU-MIMO):* When set of MIMOs (base station) communicates with single or multiple users, it is called MU-MIMO. MU-MIMO provides access to multiple users like orthogonal frequency division multiple access (OFDMA). MU-MIMO started with the fourth generation in the year 2012, with 100 Mbps data rate. Due to multi-user multiplexing, it provides direct gain. MU-MIMO is less affected by channel link loss and mutual coupling issues. MU-MIMO requires perfect CSI to provide capacity gain, multiplexing gain, and high throughput for high SNR conditions. Due to the requirement of CSI, the available channel bandwidth is completely utilized. The space division multiple access (SDMA), massive MIMO, coordinated multi-point MIMO, and ad-hoc MIMO are the distinguished forms of MU-MIMO that offer variety of services to multiple users. MU-MIMO has been integrated with 3GPP and WiMAX standards by the number of companies like Samsung, Ericsson, Nokia, and Intel, etc. MU-MIMO is suitable for less complex mobile phones with less receiving antennas. Enhanced MU-MIMO requires advanced decoding and precoding techniques. MU-MIMO is basically designed for time division duplexing (TDD) and frequency division duplexing (FDD) [37].

(iii) *Point to point MIMO* is easy to implement but severely affected by propagation conditions. Less time is required for CSI here, but the selection of the user is not flexible in point to point MIMO. Also, complex precoding, less rate gain, more error rate, and only uplink CSI are certain limitations of point to point MIMO [37].

In MU-MIMO, knowledge of uplink and downlink, CSI is required to predict the channel. MU-MIMO is less affected by environmental changes and is more flexible in user selection. Along with these, MU-MIMO provides high rate gain, less error rate, growing complexity of coding and decoding, takes sufficient time to acquire CSI, and uses simple precoding techniques [38].

3.4 Applications of MIMO

In modern scenario because of the technological boom, MIMO is considered as the key of the wireless applications in indoor and outdoor activities. MIMO has numerous applications for the following requirements:

(i) *Data rate extension:* MIMO with low bandwidth and high spectral efficiency solves the problem of high speed and low speed users. In MIMO 1 Gbps data rate is achieved with 20 MHz bandwidth only. MIMO with different modulation like binary phase shift keying (BPSK), quadrature phase shift keying (QPSK), and quadrature amplitude modulation (QAM), offers variable data rates. Data rate may also be extended further with $N_t \times N_r$ combinations.

(ii) *Range extension:* Multipath reflection, diffraction, and scattering itself are responsible for range extensions. Due to these properties, signals reach the desired object.

(iii) *Large coverage:* Increasing data rate results in coverage of large area. More and more users can be accommodated by increasing data rate. Also, the spectrum of MIMO is able to reach/cover long distances.

(iv) *Large spectral efficiency:* Single-input single-output (SISO) has very low spectral efficiency, and hence requires large bandwidth to achieve high data rate. MIMO on the other hand has very high spectral efficiency.

(v) *Bandwidth saving:* In SISO high data rate is achieved at the cost of very high bandwidth. MIMO offers high data rate at very low bandwidth. Fourth-generation long-term evolution (4G-LTE) has scalable bandwidth for various TDD and FDD uplink and downlink bands.

(vi) *Power saving:* SNR in SISO depends on the power consumption. MIMO does not consume much power as compared to SISO. Therefore it saves power for wireless applications.

(vii) *Capacity enhancement:* Capacity of MIMO is very high and is not dependent completely on the bandwidth. Capacity of MIMO linearly increases with $N_t \times N_r$. Capacity is affected by the SNR. Hence, in multipath propagation, CSI plays an important role in deciding the actual capacity of the system.

(viii) *Link reliability/improvement and dedicated services:* Even with rich multiple path propagation, MIMO ensures link reliability and dedicated links with pattern diversity and polarization diversity techniques. Low delays, low latency, and high gain are the advantages of high quality of services provided by MIMO.

(ix) *Polarization balancing:* MIMO with circularly polarized (CP) antennas ensure the reception of signals in both vertical and horizontal directions, and irrespective of polarization of the transmitter.

(x) *Wireless applications:* MIMO-LTE, WLAN-MIMO, WiMAX-MIMO, MIMO-UWB, MIMO-OFDM, MIMO-OFDMA, MIMO-SDMA, MIMO-beamforming, SU-MIMO, MU-MIMO, massive MIMO are certain applications of MIMO with the existing wireless standards. Every day, MIMO applications are becoming important due to its qualities.

(ix) *MIMO radar:* MIMO radar is the solution of poor identification capabilities of single antenna radar. In recent years, it has attracted the users for detection, estimation, and identification. Other applications of MIMO in radar are in synthetic aperture radar, and radar imaging.

(xii) *Non-wireless communication*: ITU-T G.9963 is an example of powerline communication which uses MIMO to transmit signals over multiple alternating current wires i.e. phase, neutral, and ground.

3.5 Concluding Remarks

MIMO with different functions like precoding, spatial multiplexing, and diversity coding along with their usability in wireless communication have been discussed in detail in this chapter. Similarly, different types of MIMO i.e. SU-MIMO and MU-MIMO are covered in details in this chapter. Finally, different applications of MIMO in wired and wireless are also discussed.

4

MIMO Antenna Performance Criteria

Most of the research, industry, and market have switched to the multiple-input multiple-output (MIMO) technology due to its remarkable presence in indoor and outdoor wireless needs. Most of the work on MIMO antenna systems has been reported and investigated for wireless local area network (WLAN) (2.4–2.484 GHz, 5.15–5.35 GHz, and 5.725–5.825 GHz), worldwide interoperability for microwave access (WiMAX) (2.5–2.69 GHz, 3.3–3.8 GHz, and 5.25–5.85 GHz), and Global System for Mobile Communications (GSM) (850.2–893.8 MHz, 880.0–960.0 MHz, 1710.2–1879.8 MHz, and 1850.2–1989.8 MHz) operating bands and for other systems. Existence of MIMO is an essential requirement in antenna technology.

4.1 Introduction

Multipath propagation experiences Rayleigh fading and this effect is controlled using multiple antennas at the transmitter and receiver. MIMO is used for high signal-to-noise ratio (SNR) conditions, whereas diversity is considered for low SNR conditions. Better quality of services can be achieved in non-line-of-sight (NLOS) conditions with the MIMO antennas.

The idea of MIMO started with the Shannon's capacity theorem [39] and its requirement in communication signals [40], multi-variate analysis over the gaussian channel [41], multi-channel digital transmission systems [42, 43], and directional digital transmission and reception using beamforming signal processing techniques [44, 45]. Richard Roy and Bjorn Ottersten improved the channel characteristics with space division multiple access (SDMA) by using antenna arrays at the base station [46]. Arogyaswami Paulraj and Thomas Kailath increased the data rate by intelligently processing the high data rate signals into low data rate signals. Greg Rayleigh, Gerard J. Foschini, and Emre Telatar utilized co-located antennas to improve the link. Bell laboratory was the first to introduce the spatial multiplexing to improve the performance parameters of the MIMO system [47].

Limitations of single-input single-output (SISO) antenna system roused the need of MIMO antenna systems at the base stations. SISO is very simple and cheap, but for Gigabit wireless systems, it requires a very large-frequency spectrum. Therefore, SISO, in general, is not applicable for Gbps wireless links because of SNR limits, high power and bandwidth requirements, and significant reduction in range in practical receivers [48, 49]. MIMO antenna is an essential part of wireless and mobile technologies for parameter characterization.

A comparative diagram of SISO and MIMO is shown in Fig. 4.1. Multiple antennas are required for the trans-reception to resolve the issues of SISO antenna system. Consequently, improvement in spatial multiplexing gain, channel capacity, and interference is achieved. MIMO is the linear function of channel capacity without bandwidth or power constraint [50].

MIMO solves the problem of systems having point to point communication i.e. a moving vehicle, train, basement of a building, mountains etc., where direct communication is critical. MIMO handles the problem of multi-path environment using precoding, spatial multiplexing, and diversity coding [33, 34, 51].

MIMO technique is characterized in terms of single-user, multi-user, and multi-antenna systems [35,36,52]. The cooperative MIMO and massive MIMO are the variants of multi-user MIMO [53–55].

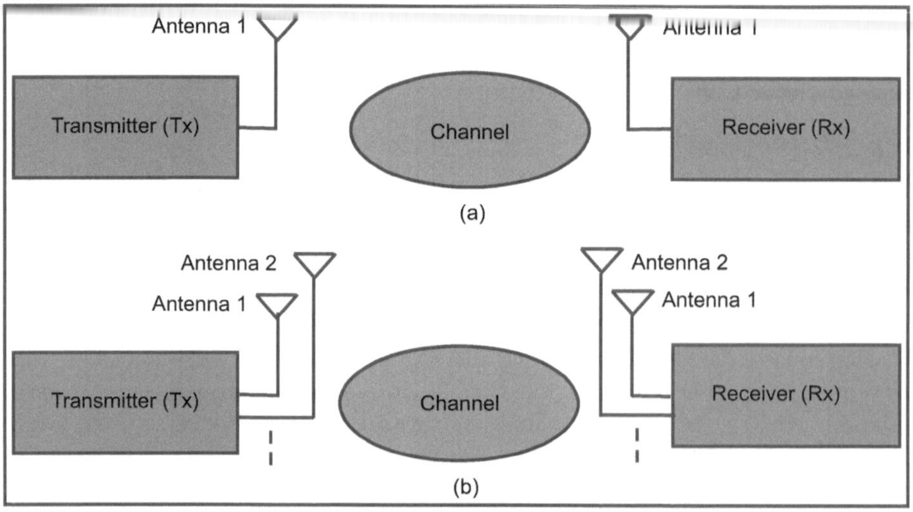

FIGURE 4.1
Block diagram of (a) SISO and (b) MIMO

Commercial wireless products i.e. mobile handset places certain bounds on space requirements. Also, the close proximity of MIMO radiating elements leads to the performance degradation due to common ground and conduction emission. This degrading effect is commonly admitted as a correlation/mutual coupling, which critically disturbs the diversity and far-field radiation characteristics. Different diversity and mutual coupling reduction techniques are used to limit the adverse effects of mutual coupling between MIMO radiators.

This chapter presents a detailed description of MIMO antenna performance parameters.

4.2 Performance Criteria of MIMO Antenna

The space reduction and mutual coupling reduction techniques play important role to control the mutual coupling between closely packed radiators. It can be understood by selecting any of the two ports of MIMO in terms of scattering parameters i.e. return-loss (S_{11}) and isolation (S_{12}). These parameters are the transformations of parameters given in the next sections.

4.2.1 Reflection Coefficient and VSWR

The reflection coefficient (Γ) is used to find the values of return-loss and isolation parameters. The relationship between Γ and VSWR is given by (4.1). The VSWR in terms of Γ can be obtained from (4.2) [4]:

$$|\Gamma| = \frac{VSWR - 1}{VSWR + 1}. \tag{4.1}$$

$$VSWR = \frac{1 + |\Gamma|}{1 - |\Gamma|}. \tag{4.2}$$

For antenna designs VSWR of 1.5:1, 2:1, and 3:1 are used. Return-loss with respect to 1.5:1 is approximately −14 dB, 2:1 is approximately −10 dB, and 3:1 is approximately −6 dB. Return-loss at any port is the function of Γ and is given by (4.3) [4]:

$$RL = -20 \log_{10} |\Gamma|. \tag{4.3}$$

4.2.2 Transmitted and Reflected Powers

The transmitted power in terms of Γ can be obtained by (4.4) [4]:

$$P_t = |\Gamma|^2. \tag{4.4}$$

The reflected power in terms of Γ can be obtained by (4.5) [4]:

$$P_{\text{refl}} = 1 - |\Gamma|^2. \tag{4.5}$$

4.2.3 Transmission Coefficient

The transmission coefficient in terms of Γ can be obtained by (4.6) [4]:

$$T = 1 + |\Gamma|. \tag{4.6}$$

4.2.4 Envelope Correlation Coefficient (ECC)

The ECC is an important diversity parameter and comes in discussion because the isolation parameters (i.e. S_{12} and S_{21}) are unable to characterize the complete antenna mutual coupling. An ECC uses all the return-loss and isolation parameters of different ports. International telecommunication union (ITU) allows for the ECC ≤ 0.5 for the mobile communication. Lower value of ECC means low coupling between radiating elements, while higher value of it shows the negative impact. An ECC is given by (4.7) [56]:

$$|\rho_e(i,j,N)| = \frac{\left|\sum_{n=1}^{N} S_{i,n}^{*} S_{n,j}\right|}{\sqrt{\left|\prod_{k(=i,j)} \left[1 - \sum_{n=1}^{N} S_{i,n}^{*} S_{n,k}\right]\right|}}, \tag{4.7}$$

where the values of $i = 1$ to n, $j = 1$ to n, n = considered first port number, k = considered next port number, and N = number of radiating elements.

For higher-frequency operations like 5G, ECC can be obtained using far-field parameters of the radiating elements. Let $\bar{F}_1(\theta, \phi)$ and $\bar{F}_2(\theta, \phi)$ are the field patterns of two radiating elements with respect to the θ and ϕ components, then ECC can be given by equation (4.8) [56].

$$\rho_e = \frac{\left|\iint_{4\pi} \left[\bar{F}_1(\theta,\phi) \bullet \bar{F}_2(\theta,\phi)\right]\right|^2}{\iint_{4\pi} \left|\bar{F}_1(\theta,\phi)\right|^2 d\Omega \iint_{4\pi} \left|\bar{F}_2(\theta,\phi)\right|^2 d\Omega}, \tag{4.8}$$

where $\bullet$ is the hermitian product operator and Ω is the solid angle.

4.2.5 Total Active Reflection Coefficient (TARC)

Similarly, TARC validates the S parameters for the diversity performance by involving random signals and their phase angles at diagonal/adjacent ports. It verifies the nature of S parameters i.e. S_{11}, S_{12}, and S_{13}, etc., for particular phase combinations between ports. Let the excitation vector a_i and scattering vector b_i are gaussian random variables (RVs) that are independent and identically distributed (iid). The equation of TARC can be written as (4.9) [56]:

$$\Gamma_a^t = \frac{\sqrt{\sum_{i=1}^{N} |b_i|^2}}{\sqrt{\sum_{i=1}^{N} |a_i|^2}}. \tag{4.9}$$

The ratio of b and a for a particular port results in the return-loss/isolation parameter (S_{ii} or S_{ij}). The above equation of TARC can also be written for adjacent/diagonal radiating ports as (4.10) [56]:

$$\Gamma_a^t = \frac{\sqrt{|(S_{ii} + S_{ij} * e^{j\theta})|^2 + |(S_{ji} + S_{jj} * e^{j\theta})|^2}}{\sqrt{N}}, \tag{4.10}$$

where θ is the phase angle between adjacent/diagonal ports. Different values of θ are selected at different ports. In that case, θ will be $\theta 1$, $\theta 2$, and so on.

4.2.6 Channel Capacity

Channel capacity is another diversity parameter which depends on the scattering parameters of the MIMO antenna. Let us consider a multi-path channel having multiple transmitting antennas (N_t) on one side and receiving antennas (N_r) on the other side. Formula of channel capacity of MIMO can be given as (4.11) [56]:

$$C = \log_2 \left(I_{Nr} + \frac{P}{NN_t} HH^* \right), \tag{4.11}$$

where I is the identity matrix, P is the signal power, N is the noise power (capacity of SISO depends on $C = B \log_2(1 + \frac{P}{N})$), $*$ denotes a complex conjugate transpose operation, H is the channel matrix and its transpose is H^*, and B is the bandwidth. Response H depends on the correlation/mutual coupling between radiators, which is the function of channel parameters. The appropriate knowledge of the channel and its parameters enables to find the value of H. The channel matrix for two-port MIMO is given as (4.12) [56]:

$$H = \begin{bmatrix} h_{11} & h_{12} \\ h_{21} & h_{22} \end{bmatrix}, \tag{4.12}$$

where h_{11}, h_{12}, h_{21}, and h_{22} are the channel parameters to decide the state of the channel.

The capacity of MIMO can also be obtained in terms of a number of streams and capacity of SISO and is given by (4.13) [56]. Let the number of streams are M.

$$C = MB \log_2 \left(1 + \frac{P}{N} \right). \tag{4.13}$$

4.2.7 Mean Effective Gain (MEG)

MEG provides information about gain in NLOS channel conditions. MEG is measured in an anechoic chamber for the far-field characterization. MEG includes efficiency and power patterns of the MIMO antenna. Let the cross-polarization ratio of the incident field is denoted by XPR (also known as cross-polarization discrimination (XPD)). The probability distribution function (pdf) of the incident waves are $p_{\theta j}$ and $p_{\phi j}$ for θ and ϕ components, and

Ω is the solid angle. The power gains of 2×2 MIMO antenna elements are $G_{\theta j}$ and $G_{\phi j}$. The equation of MEG in 2D plane is given by (4.14) [57]:

$$MEG_j = \oint \left(\frac{XPR}{1+XPR} p_{\theta j}(\Omega) G_{\theta j}(\Omega) \; + \; \frac{1}{1+XPR} p_{\phi j}(\Omega) G_{\phi j}(\Omega) \right). \tag{4.14}$$

The above equation of MEG can also be written by (4.15) [57]:

$$MEG_j = \frac{1}{2\pi} \int_0^{2\pi} \left[\frac{XPR}{1+XPR} G_{\theta j} \left(\frac{\pi}{2}, \phi \right) + \frac{1}{1+XPR} G_{\phi j} \left(\frac{\pi}{2}, \phi \right) \right] d\phi. \tag{4.15}$$

Other equations of MEG may be given by (4.16) and (4.17). Let MEG_1 and MEG_2 are due to port 1 and 2 (in case of two ports, and can be extended for a number of ports) [57].

$$MEG_1 = 0.5\eta_{1,\mathrm{rad}} = 0.5 \left[1 - |S_{11}|^2 - |S_{12}|^2 \right], \tag{4.16}$$

$$MEG_2 = 0.5\eta_{2,\mathrm{rad}} = 0.5 \left[1 - |S_{12}|^2 - |S_{22}|^2 \right], \tag{4.17}$$

where $\eta_{1,rad}$ and $\eta_{2,rad}$ are the radiation efficiencies at ports 1 and 2.

MEG with 100% efficiency is given by −3 dB. For MIMO, each element should have an equal value of MEGs to get optimum diversity performance. The considered value of $XPR = 0$ dB is for the outdoor environment and $XPR = 6$ dB for the indoor environment. MEG includes the effects of $G_{\theta j}$ and $G_{\phi j}$, which are the functions of mutual coupling between MIMO antenna elements.

4.2.8 Spectral Efficiency

MIMO serves better results in respect of spectral efficiency. The spectral efficiency of fourth generation long-term evolution (4G-LTE) is 4.08 bits/second/Hz for SISO and 16.32 bits/second/Hz for 4×4 MIMO technology. Similarly, the spectral efficiency of 4G-LTE advanced is 3.75 bits/second/Hz for SISO and 30.0 bits/second/Hz for 8×8 MIMO technology [50].

Spectral efficiency (bits/second/Hz) is calculated as per equation (4.18):

$$\eta = \frac{R}{B}, \tag{4.18}$$

where η is spectral efficiency, R is data rate, and B is the bandwidth of channel.

4.2.9 MIMO Mode

The mode of the MIMO antenna return-loss can be obtained by (4.19) [50]:

$$f_r = \frac{1}{2\pi\sqrt{\mu\varepsilon}}\sqrt{\left(\frac{m\pi}{h}\right)^2 + \left(\frac{n\pi}{L}\right)^2 + \left(\frac{p\pi}{W}\right)^2}, \tag{4.19}$$

where h is the substrate height, L is the length of patch, and W is the width of patch.

4.3 Concluding Remarks

MIMO antenna provides better quality of services and performance parameters i.e. diversity and far-field, etc., in comparison with SISO. MIMO antennas are the indistinguishable part of the present and future generations of wireless communications. They are implanted at the physical layer of internet protocol (IP) to establish unending signaling link, provide low delays, high capacity, and high data rate etc. The performance parameters of designed MIMO antenna unveil the diversity and far-field behavior for the compactness and low correlation among radiators.

5

5G Massive MIMO Technology

The most dominating technology in fifth generation is the massive MIMO. Massive MIMO uses all the benefits of conventional MIMO on a large scale. Number of antenna arrays in massive MIMO serves a large number of terminals simultaneously. It provides an infrastructure and clouds to interact with things and people via internet of things. The spatial multiplexing (SM) in massive MIMO and channel state information by base stations leads to the perfect uplink and downlink operation. Aggressive SM tends to enhance the capacity of massive MIMO. A large number of antenna arrays focuses the radiated energy in all possible locations of terminals constructively to enhance the radiated energy efficiency and destructively elsewhere to suppress interference. The degree of freedom in massive MIMO shapes the signal for hardware. Due to a large number of antennas, fading dips are avoided and latency can be controlled. Joint operation of channel state information with pilot signaling in uplink and decoding techniques diminishes the signal problem due to intended jammers in massive MIMO. Proper planning of precoding and decoding strategies in massive MIMO helps to approach the Shannon limit. The increase in the number of base station radiators more than the users, as well as the channel knowledge of base station about downlink, are the other approaches to achieve the Shannon limit.

5.1 Introduction

The fourth-generation (4G) technology was commercially deployed in Oslo, Norway, Stockholm, and Sweden in 2009, and since then was spread to the rest of the world. In March 2008, ITU specified that the maximum peak speed of the technology should be 100.0 Mbps for low mobility users like a person travelling/walking on road, and 1.0 Gbps for high mobility users like persons travelling in car/train/aeroplane etc. The 4G technology uses frequencies in the range of 700.0 MHz, 850.0 MHz, 950.0 MHz, 1800.0 MHz, 1900.0 MHz, 2100.0 MHz, 2300.0 MHz, 2600.0 MHz, 3500.0 MHz, etc. [58, 59].

The 4G technology was designed to work over internet protocol (IP) packet switching network and is able to dynamically share and use the network resources to support multiple users per cell. The 4G technology uses a scalable channel bandwidth of 5.0 to 20.0 MHz, optionally up to 40.0 MHz and has peak spectral efficiency of 15.0 bits/s/Hz in the downlink and 6.75 bits/s/Hz in the uplink. Also, the system spectral efficiency in indoor cases is 3.0 bits/s/Hz/cell for downlink and 2.25 bits/s/Hz/cell for uplink and having smooth handovers across heterogeneous networks. The 4G has some other features like use of MIMO, turbo principle error-correcting codes, channel-dependent scheduling, link adaptation to spread a boon in the market for wireless users. Major advantages of 4G technology are that it has low latency data transmission, variable data speed from 100.0–300.0 Mbps, has scalable bandwidth, provides uninterrupted connectivity especially for video chats and conferences, better coverage, more secure and safer than previous technologies, and has affordable cost-effective/pricing schemes [60].

Apart from its pros, there were some cons which led scientists to switch to better technologies. Although, it offers uninterrupted connectivity, but to some regions, its connectivity gets limited, many users are annoyed by its glitches and bugs, and it consumes lots of battery power. Those who cannot access to 4G, have to switch to 3G/Wi-Fi while paying for the same amount as specified by 4G network plan, also users are forced to buy new devices to access 4G technology. Although, 4G LTE promised many features like serving 1 million base stations for outdoor and indoor applications, still it failed to satisfy the exponentially increasing demands of mobile and wireless users. Many researchers and mobile/wireless industries did their level best to increase data rate for worst case conditions, antenna gain, etc., but failed to do so due to growing user needs [61, 62].

The fifth generation (5G) is built on IEEE 802.11 protocol and aimed to speed up its data rate three times that of 4G. The mobile devices used with 4G can be modified to accommodate 5G antennas. The mobile companies Xiomi and Samsung are working on the world's first 4G and 5G enabled mobile phones, which will work on 28 GHz (millimeter wave) and 3.5 GHz for 5G broadcasting, and 2.5 GHz 4G bands. Also, the 5G antenna can be assembled between long-term evolution (LTE) and Wi-Fi antennas of 4G mobile phones [33, 34, 63].

5G which is promising to deliver upto 1000 times as much data as of today's networks, will also consume upto 1000 times as much energy as now. This consumption is becoming a hot topic to concern about. 5G small cell base stations are proposed to consume more than 50% of energy and can approach upto 800 watts with massive MIMO for high volume traffic. Hence, computation of power can play a vital role in energy efficiency of small cell networks. As computation power increases with the heavy traffic, transmission power reduces. Thus, there is a trade-off between the two. On comparing these two, former is small compared to later, therefore, the energy efficiency investigation of small cell networks focuses on the optimization of transmission power at the base station. In this technology, power can be saved by BS sleeping scheme, where the radio frequency (RF)s and transmitters of BS are closed during no traffic. As the number of antennas and bandwidths are increased, computation power also increases. Thus in 5G, computation power will play a more important role than transmission power, therefore both the powers should be considered while optimizing energy efficiency in 5G small cell networks [64].

The lower spectrum of 5G bands are covered under the 3.3–4.2 GHz and 4.4–4.99 GHz. The higher-frequency spectrum for 5G band covers frequencies like 24.25–27.5 GHz, 26.5–27.5 GHz, 26.5–29.5 GHz, 27.5–28.28 GHz, 27.5–28.35 GHz, 37.0–40.0 GHz, and 37.0–43.5 GHz which are deployed in countries like China, Korea, Japan, etc. Frequencies above 60.0 GHz are 53.3–66.5 GHz, 55.4–66.6 GHz, 56.6–64.8 GHz, 57.0–64.0 GHz, and 57.0–65.0 GHz.

This chapter presents a detailed description of massive MIMO communication system.

5.2 Massive MIMO

Massive MIMO is a system of 5G technology of $\geq$ 100 radiators to serve a very large number of terminals and is a broadband technology. Massive MIMO research is going on to provide 10 Gbps data rate for higher generations of wireless communication systems till the year 2020. Massive MIMO benefits can be increased by combining it with beamforming i.e. maximum ratio transmission, maximum ratio combining, and zero-forcing techniques. Massive MIMO has integrated with LTE and IEEE 802.11 wireless technologies. Channel estimation is done in massive MIMO with orthogonal pilot sequences. Hence, special pilot assignment strategies must be adopted to solve the problem of pilot contamination due to co-channel cells. Such an approach results in very high channel capacity. Massive MIMO is designed for time division duplexing (TDD) only. Massive MIMO benefits can also be considered in indoor and outdoor activities. Massive MIMO provides guaranteed coverage and smartly handles the mobility of users in outdoor environments. There is no need of putting base stations inside the building or premises for indoor activities. Some companies like Ericsson, Huawei, and ZTE have tested massive MIMO using 64 to 128 antennas.

5.3 Channel Estimation in Massive MIMO

Frequency responses of the propagation channels are measured in massive MIMO. Transmission of training signals from the base stations or users is responded by the receivers to provide the channel state response. In limited time and before the movement of the users, the channel estimation is utilized. Massive MIMO simultaneously controls the training and transmission of data in a slot. The movement of the user is the fraction of the wavelength of slot duration. Instead of assuming the channel, massive MIMO measures the channel information directly. This property makes massive MIMO a scalable technology.

In TDD, users/terminals transmit pilot sequences for training in each Nyquist interval. For training, the slot duration is expanded till half because in TDD shorter slots have higher mobility. Only uplink channel state information is required in TDD. The uplink and downlink channels are reciprocal in TDD. On the other hand, frequency division duplexing (FDD) requires more time than TDD for training. FDD uses different uplink and downlink bands, hence different uplink and downlink channel state information are obtained. FDD is limited to a small number of antenna systems while TDD provides a drastic advantage for a large number of antenna systems i.e. scalability for active users.

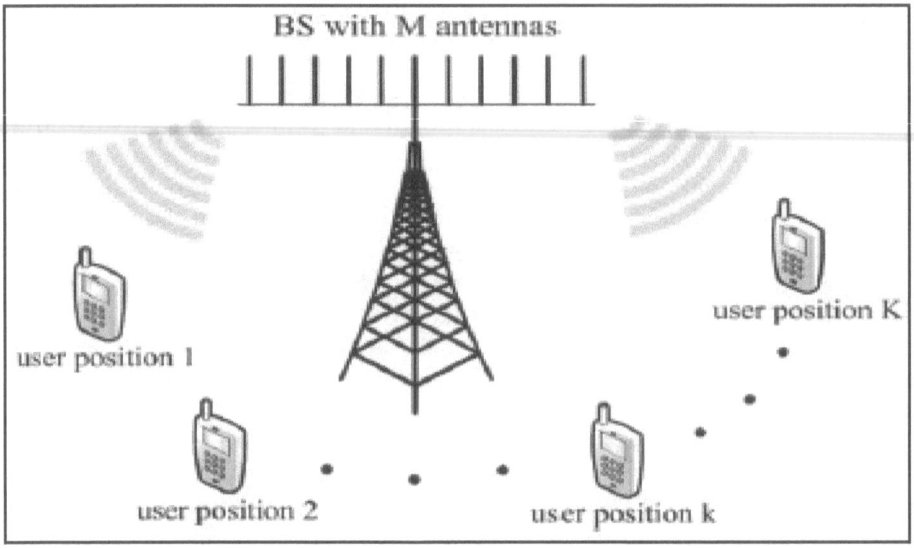

FIGURE 5.1
Massive MIMO system [65]. [From: Xudong Cheng, Yejun He, Li Zhang, and Jian Qiao, "Channel modeling and analysis for multipolarized massive MIMO systems," International Journal of Communication Systems, pp. 1–16, April 2018. Reproduced courtesy of the John Wiley & Sons, Ltd.]

The overheads in TDD and FDD can be seen using an example. Let the length of coherence block be $\tau = B_c T_c$ symbols. Per coherence block there are K uplink symbols in TDD and downlink poilets are eliminated due to channel hardening. Whereas, FDD uses M poilet symbols per coherence block in downlink band and K poilet symbols with the feedback of M channels per terminal in the uplink band. Therefore, $M + K$ uplink symbols are required in FDD. Figure 5.1 shows the operation of massive MIMO with each base station having a set of M transmitting antennas and K mobile terminals [65].

The operation of the TDD and FDD with (M, K) points can be understood with respect to τ. For $M = 100$ and $K = 25$, overheads in both the TDD and FDD can vary. The overhead in FDD is very high for high frequencies or high mobility cases. The critical limit is imposed for high frequency/mobility cases in FDD (i.e. for $\tau = 200$). It is examined that the overhead is dependent on the number of terminals in TDD and independent on antennas for uplink channel [66].

5.4 Spatial Diversity/Multiplexing

Spatial diversity is the concept of increasing the reliability of the continuous signaling in MIMO technology. In spatial diversity, each signal is sent via multiple paths from the transmitter. Spatial multiplexing is the complex form of spatial diversity, where multiple messages are transmitted by separating them in space to avoid interference in the same frequency band. Hence, more data transmission through multiple paths between transmitter and receiver is carried out. The spatial multiplexing is one of the key techniques to enhance the wireless network performance in present MIMO communication.

5.5 Beamforming

By the use of advanced antenna technologies both at the transmitter and receiver, beams are created instead of broadcasting the signal in a wide area. The transmit and receive beamforming techniques are used at the base stations or mobile stations to maximize received signal strength of the user and also to control/minimize the interference from nearby users. Massive MIMO works on the beamforming concept to provide better coverage of the user by concentrating signals in a particular direction. In beamforming, the user is continuously tracked by the base station due to feedback of mobile at any point in vertical and horizontal directions. This approach of searching the user by vertical beam and horizontal beam in a direction creates 3D beamforming, which in turn enhances data rate and capacity. Due to the focused narrow beams, interference between different beams can be controlled in various directions.

The beamforming is used for millimeter (mm) waves to cover the unused frequency spectrum at high frequencies. Because the frequency spectrum below the 5 GHz is licensed for urban communication. At higher frequencies, higher bandwidths are available. The problem of poor propagation characteristics at high frequencies can be controlled by high gain and compact directional antennas at mobile terminals. Instead of using a fixed narrow beam for mobile terminals, beamforming is the best solution [67].

5.5.1 Types of Beamforming Techniques

Various beamforming techniques are discussed in this section. These are as follows:

(1) *Wideband and narrowband beamforming:* These are decided by the signal bandwidth. The combination of received signal arrays results in narrowband beamforming. Similarly, wideband signals form the wideband beamforming. In wideband beamforming, finite impulse response (FIR) filter, infinite impulse response (IIR) filter, and delay lines are used to control the widebeam operation [68].

(2) *Switched and adaptive beamforming:* The switched beamforming is a fixed beamforming and depends on the switching network like Butler matrix to cover a particular terminal. There is a possibility of uncertainty to attain the desired direction and beamforming is unable to serve multiple mobile terminals [69].

    Adaptive beamforming is able to cover each mobile terminal by forming separate beams. This is possible by utilizing adaptive signal processors to control amplitude and phase in view of the detected mobile terminal. The main lobe of beams is selected in the preferred direction of the mobile terminal and nulls in other direction to reduce interference. The adaptive beamforming is more difficult to implement than switched beamforming. Still, adaptive beamforming is preferred in massive MIMO [70].

    Adaptive beamforming performs its operation by using blind and non-blind algorithms. The least mean square (LMS), recursive mean square (RLS), sample matrix inversion (SMI), and conjugate gradient algorithm (CGA) are non-blind adaptive algorithms. The constant modulus algorithm (CMA) and least square CMA (LSCMA) are the blind adaptive algorithms. No reference signal is implemented in blind adaptive algorithms, therefore array weights are not controlled [71].

    The digital signal processor in adaptive beamforming controls the shape and direction of radiation pattern (beam) as per received signal. Hence, maximum gain is achieved due to noise and interference minimization in an intended direction by adaptive array beamforming. Adaptive beamforming is used in smart antennas and also in digital beamforming.

    Based on the null steering, some other adaptive beamforming approaches like linearly constrained minimum variance (LCMV) and minimum variance distortionless response (MVDR) are also used to improve the signal to interference and noise ratio (SINR) by adapting nulls in interference positions. MVDR is preferred in massive MIMO. LCMV and MVDR null approaches utilize the azimuthal and elevation angles to enhance the performance of the adaptive system by interference reduction and accuracy enhancement [72].

(3) *Analogue, digital, and hybrid beamforming:* Analogue beamforming is an old approach and works on the adjustment of the phase of each transmitted signal using fixed phase shifters. Beam steering is performed here using RF switch. Modern analogue beamformers are able to provide continuous beamforming. Null controlling is not so easy in analogue beamforming.

Digital beamforming controls radiation patterns of antennas, involves the direction of arrival (DOA) of signals, and SINR by adaptive steering both the beam and null. Due to complex operation, digital beamforming is not used in massive MIMO [73].

Analogue beamforming uses expensive phase shifters and is costlier than digital beamforming. Also, the phase shifters are not flexible in amplitude in analogue beamforming, thus the performance is poor here.

Hybrid beamforming combines the effects of analogue and digital beamforming techniques to achieve better performance. Here, analogue beamforming controls the number of analog to digital converters (ADCs) and digital to analog converters (DACs) at RF, and digital beamforming utilizes baseband signals. Hence, the joint operation of these two techniques provides the cost effective solutions and is mostly preferred in massive MIMO [74].

5.6 Advantages of Massive MIMO

Massive MIMO is the key technology of the 5G with more than 100 antennas at both transmitting/receiving base stations to provide the better services to end user. It has the following advantages:

1. *Capacity enhancement:* Higher capacity (10-fold or more) is expected in 5G wireless communication due to the massive MIMO without any additional spectrum. Total available data resources are utilized by the maximum end users for specific requirement due to massive MIMO. Massive MIMO is the key resource for the 5G higher data rates.

2. *Large coverage:* Users can experience better coverage inside as well as at the edges of a cell. Hence higher data rates are achieved everywhere. Use of beamforming creates a most suitable beam for the stationary and moving users in massive MIMO.

3. *User experience:* Large files/data can be uploaded/downloaded using massive MIMO in 5G due to large capacity and large coverage. So, better user experiance is expected in massive MIMO.

4. *Link reliability:* More number of possible paths are created in massive MIMO between transmitter and receiver due to a large number of antennas at both ends. Hence, reliability is enhanced here.

5. *Resistance to interference and jamming:* Multiple paths and narrow beams solve the problem of interference and jamming in case of massive MIMO.

6. *Targetted use of spectrum:* Massive MIMO with beamforming utilizes available spectrum smartly even in dense locations.

7. *Low power consumption:* The base stations with beamforming require low power to cover the intended user, thus reducing the overall cost of the network. Only the number of working antennas are required in a particular beam at the base station. Therefore, energy efficiency is improved.

8. *System security:* An intended user is covered using the beam steering transmitted signal. Therefore, signal theft is minimum in a massive MIMO system as compared to the conventional system.

5.7 Concluding Remarks

In this chapter the massive MIMO antenna system has been discussed in detail. Various frequency bands in 5G, channel estimation, overheads in TDD/FDD, spatial diversity/multiplexing techniques, and different types of beamforming techniques have been discussed. These all the techniques in massive MIMO are used to enhance the performance of the channel link and to provide the user with the maximum data rate, throughput, capacity, and minimum interference. Various advantages of the massive MIMO have also been covered here.

6

Mutual Coupling Reduction Techniques in MIMO Designs: An In-depth Survey

Variety of mobile applications have increased the need for Gigabit data rates to accommodate every class of users (low mobility and high mobility both). To accommodate such class of users, a single antenna is replaced by the MIMO antenna and is implanted inside mobile devices/phones. Due to the use of MIMO, high quality of services like un-ending signaling, high data rate, high capacity, and high spectral efficiency are achieved. However, compactness is the main issue which leads to the mutual coupling among the MIMO radiators at the physical layer. Variety of solutions are given to solve the issue of correlation. Various diversity and mutual coupling reduction techniques are given in this chapter.

6.1 Introduction

Presently available mobile devices require the MIMO antennas for high capacity and high data rate applications. However, the close proximity of radiators in MIMO leads to the degradation in antenna performance parameters. Consequently, far-field gain, efficiency, correlation, and current distribution require high attention for the desired quality of user services. These effects are controlled using the space, pattern, and polarization diversity techniques. Along with these, other techniques are also required to control the effects of correlation.

In this chapter a thorough description of diversity techniques, decorrelation (isolation enhancement) techniques, and antenna miniaturization techniques are discussed based on their performance and utilization.

6.2 Diversity Techniques

Various types of diversity techniques used in MIMO antenna designs are discussed in details here.

6.2.1 Space Diversity

Two closely coupled antenna radiators can be de-correlated using the space variation. The space variation is the key of this approach to achieve low mutual coupling. All types of MIMO antennas like conventional and non-conventional shaped monopole, dipole, fractal, planar inverted frequency antenna (PIFA), etc, use the space diversity. Large variation leads to the desired performance, but at the cost of drastic increase in overall MIMO dimension [75]. The coupling (radiation emission) and common conductor (ground) emission are the sources of mutual coupling in closely packed radiators. The behavior of correlation coefficient reveals the nature of a designed MIMO. The mutual coupling is studied in terms of ground and patches, separation and orientation of radiators [76].

The overall size with the proper treatment of the antenna parameters places the upper bound on the space requirement. Position and orientation of the radiator like diagonal placement, anti-face (180° face shifted) or face to face, etc., is the prime requirement to enhance isolation and performance parameters of MIMO antenna [77, 78]. The direction of the current flow is changed in space orientation technique to control the correlation effects. Variation in WLAN/WiMAX/LTE band MIMO antenna bandwidth on FR-4 dielectric substrate of size 50×100 mm^2 was achieved with space diversity [70]. The mutual coupling also depends on the type and shape of the MIMO antenna radiators [80].

6.2.2 Polarization Diversity

Space diversity shows better utilization at the base stations, due to large space availability. However, strict space constraint is required for mobility and portability of present wireless and mobile entities. Another solution, in terms of polarization diversity, has received the high attention of the mobile and wireless devices. The placement of MIMO radiators on vertical and horizontal diversity branches results in large capacity due to the high quality of services and compactness of the designed antenna [81]. The placement of radiators in the orthogonal position leads to lower installation costs and low mutual coupling [82].

The combined effect of pattern and polarization diversity leads to enhancement in link utilization and low correlation among radiators of MIMO i.e. 120° apart radiating patches. Consequently, widening in lobe beamwidth and higher diversity gains are achievable [83]. MIMO with triangular patches on FR-4 dielectric substrate of size 90×120 mm^2 are arranged in polarization diversity for size reduction and isolation enhancement [84, 85]. The change in field orientation of radiators eliminates most of undesired coupled energy in polarization diversity [86] and is used for spectrum utilization in telecommunication radios [87].

A design with Taconic (ORCER) dielectric substrate (thickness 1.52 mm) of size 70.11 × 70.11 mm^2 and with orthogonal slots, and metal via to short ground, was designed to improve performance parameters [88]. The polarization diversity with a slotted strip in multi-layer MIMO [89], and crossed exponentially tapered slots [90], enhance the isolation among radiators. MIMO designs with polarization diversity achieve self-balancing radiation patterns. A fabricated MIMO antenna on FR-4 dielectric substrate (thickness 1.6 mm) of diameter 110 mm was designed with such concept [91]. Also, the port decorrelation [92], reduction in coupling current [93], efficiency improvement [94], and desired bandwidth with in-band isolation responses [95] are the merits of polarization diversity technique. More than 15 dB of isolation can be achieved in this approach with proper planning [96].

6.2.3 Pattern Diversity

The quality of services in multi-path environments ensures the good radio link reliability due to pattern diversity by creating far-field radiation patterns at different angles. Also, the symmetry/asymmetry of radiation patterns depends on the placement and orientation of radiating patches. The complementary symmetry in radiation patterns may amount to the symmetric isolation and return-loss responses at ports. Bandwidth enhancement and reconfigurability in multi-mode and multi-layered structures may be achieved due to stepped profile and pattern diversity [97, 98]. Ground current altering can be done by pattern reconfigurability. The curved slots [99] and I-shaped radiator [100] are utilized to implement pattern diversity. The shape of the radiating element is responsible for the amount of mutual coupling in this approach.

6.3 Mutual Coupling Reduction Techniques

Closely packed MIMO radiators with different mutual coupling reduction techniques may have variation in the amount of isolation produced. These are discussed in details here.

6.3.1 Parasitic Element/Structure Approach

The parasitic element is a non-contacting metallic/non-metallic structure between the MIMO antenna radiators, and is placed with keen observation of the field current to enhance return-loss bandwidth, isolation, and far-field characteristics [101]. Parasitic elements produce surface currents in the existing coupling path to control ohmic losses [102]. MIMO with dumb-bell shaped parasitic element on FR-4 dielectric substrate (thickness 1.6 mm) of size 40 × 40 mm^2 is shown in Fig. 6.1 [103].

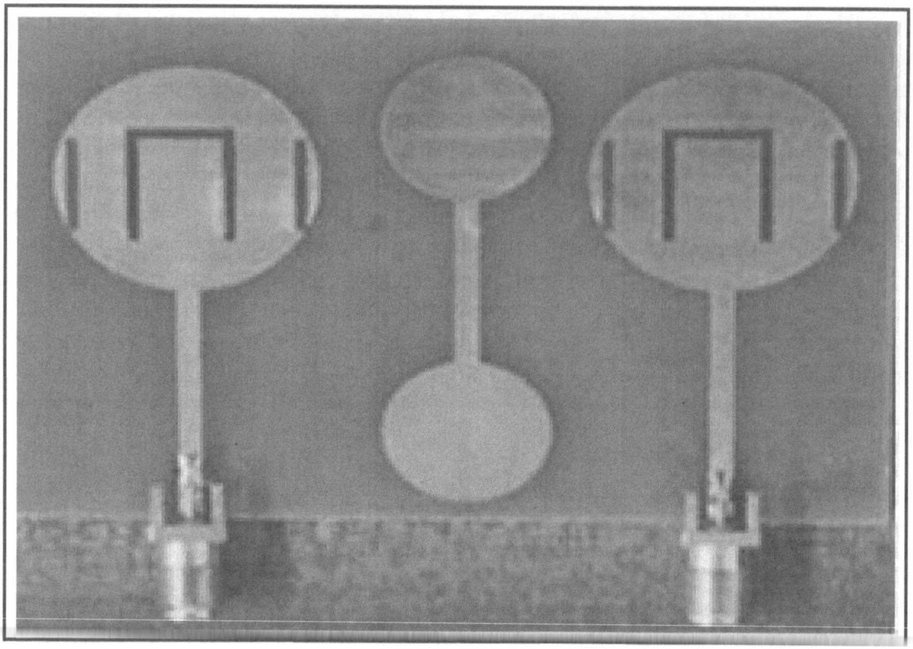

FIGURE 6.1
Parasitic element between MIMO radiators [103]. [From: Kommana Vasu Babu, and Bhuma Anuradha, "Analysis of multi-band circle MIMO antenna design for C-band applications," Progress in Electromagnetics Research C, vol. 91, pp. 185–196, March 2019. Reproduced courtesy of the Electromagnetics Academy.]

Similarly, the parasitic element cancels out the field by reverse coupling and limits the effect of correlation at ports. The design was fabricated on FR-4 dielectric substrate (thickness 0.8 mm) of size 60×95 mm^2 [104]. The position and placement of the parasitic element lead to critical analysis [105], and it forms the inductive loading equivalence to cancel mutual coupling in the desired frequency band. Better than 15 dB isolation is seen in many MIMO antenna designs with parasitic elements.

6.3.2 Neutralization Line Approach

To control the effects of mutual coupling, a line called the neutralization line is connected to feeding and/or shorting points. It results in inductive loading and forms the inductor-capacitor equivalent circuit due to EM field by reverting the coupling current. It is a port de-coupling approach which produces open circuit to obtain the in-band isolation at the desired impedance bandwidth [106, 107]. MIMO antenna on FR-4 dielectric substrate (thickness 1.0 mm) of size 30×65 mm^2 with neutralization line was implemented with short-circuited monopoles to achieve better isolation between ports [108]. MIMO antennas with shorting posts for desired far-field radiation patterns use the neutralization lines [109]. However, searching a low-impedance path in compact designs for the neutralization line needs the critical analysis, also the amount of isolation is design dependent.

6.3.3 Slit and Slot Etching Approach

A slot is a structure which helps in improving the compactness, efficiency, bandwidth, isolation, as well as creates multi-band operation of MIMO [110–112]. Similarly, a slit forms the capacitive body and controls the effect of correlation at ports of MIMO [113]. Both slits and slots are the slow-wave structures and form back lobes. Some designs like PIFA antenna with slots and slits and with $\lambda/4$ wavelength resonator control radiation in back direction and offer compactness, improve efficiency but require matching circuits also to control correlation effects [114]. The increase in patch height helps to improve isolation. Antenna with such concept was designed on FR-4 dielectric substrate (thickness 0.8 mm) of size 60×80 mm^2 [115]. The slits also improve the bandwidth due to capacitive property [116].

A cavity is introduced in exponentially tapered slot and intersecting square slot to control the back lobe and limit the coupling effects [117]. The cavity backed slot array antenna is the source of re-radiation of electromagnetic energy and alters the radiation pattern [118]. The defected ground structure [119] and H-shaped slot [120] also limit the adverse effects of mutual coupling.

An electromagnetically excited loop with U-shaped slot was designed on CEM-1 S3110 dielectric substrate (thickness 0.8 mm) of size 50×63 mm^2 to control the mutual coupling and also extends the bandwidth [121]. Similarly, a MIMO antenna with a T-shaped slot and a chip capacitor was designed for band elimination and isolation enhancement, on FR-4 dielectric substrate (thickness 0.8 mm) of size 55×90 mm^2 [122]. For isolation improvement and multi-band operation, slots and ground branches were designed on FR-4 dielectric substrate (thickness 0.8 mm) of size 60×120 mm^2 [123]. The isolation at ports is design dependent and it may be 20 dB.

6.3.4 Coupling/Decoupling Structure Approach

A microstrip line of $\lambda/4$ wavelength or sub-multiple of it may be used for decoupling/de-scattering to control the reactive term at some locations in design [124]. An indirectly coupled coupling element was designed on FR-4 dielectric substrate (thickness 1.6 mm) of size 20×40 mm^2 to introduce extra current path for de-correlation between antennas [125]. Decoupling structure converts the trans-admittance into pure imaginary part [126]. Ring hybrid decoupling structure can be used for de-correlation [127]. Antenna characteristics may be improved by combining decoupling and matching networks [128, 129]. A directly connected decoupling structure creates the dedicated far-field radiation patterns and also protects the load from coupling current. MIMO antenna with such concept was designed on FR-4 dielectric substrate (thickness 1.56 mm) of size 20×60 mm^2 [130].

The U-section [131] and neutralization tuner [132] type decouplers improve the performance parameters of MIMO designs. MIMO antenna with decoupling network, microstrip line, and matching stub is used for proper impedance matching and desired outcome in terms of antenna performance parameters [133]. The diversity parameters can be improved with different shapes of decoupling networks [134–136]. Amount of coupling is again design-dependent and may be approximately 15 dB in this approach.

6.3.5 Metamaterials Approach

Various types of open-ended metamaterial inspired EBG structures form the L-C equivalent resonant circuit to control mutual coupling among radiating elements. These structures improve the diversity gain, matching, and prevent the shape of far-field patterns. Multiple layers of EBG offers high isolation in the band [137–139]. Split ring resonator (SRR) and complementary split ring resonator (CSRR) are the open-ended ring-pair structures to control the effect of mutual coupling [140–148]. MIMO antenna on FR-4 dielectric substrate (thickness 0.8 mm) of size 50×100 mm^2 uses the concept of CSRR [149]. The coupled loop may behave like a SRR/CSRR and enhances isolation among radiators [150, 151]. Babinet's principle helps in field analysis of SRR/CSRR. MIMO with such concept was designed on FR-4 dielectric substrate (thickness 0.8 mm) of size 70×100 mm^2 and is shown in Fig. 6.2 [152].

MIMO with complementary pattern approach using composite right/left handed (CRLH) metamaterial transmission line improves compactness and isolation [153]. A slotted meander line resonator (SMLR) produces notch band characteristics to limit the correlation between radiators. MIMO with SMLR was designed on FR-4 dielectric substrate (thickness 1.6 mm) of size 45×54 mm^2 [154]. A modified serpentine structure (MSS) controls the surface currents [155]. MIMO antenna with SRR offers multi-band operation, compactness, and less mutual coupling [156]. The port-to-port isolation may be 15 dB in this approach.

6.3.6 Shorting Pins/Posts

The shorting pins/posts are the type of inductive loading and form L-C equivalent circuit. Sometimes cross-polarization distorts the gain patterns of the antennas due to shorting pins/posts. The cross-polarization is controllable by the use of meandering probe feeding structure. Consequently, high bandwidth is achieved [157, 158]. Specific design parameters may be obtained with shorting pins/posts [159]. Portable wireless devices with multi-branch MIMO and shorting strip improve isolation by diverting a certain amount of

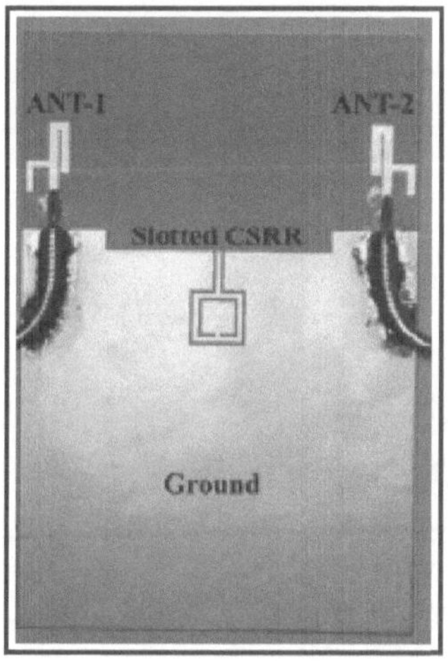

FIGURE 6.2
MIMO antenna with slotted CSRR [152]. [From: D. G. Yang, D. O. Kim, and C. Y. Kim, "Design of a dual band MIMO monopole antenna with high isolation using slotted CSRR for WLAN," Microwave and Optical Technology Letters, vol. 56, no. 10, pp. 2252–2257, October 2014. Reproduced courtesy of the John Wiley & Sons, Ltd.]

current [160]. Shorting strip creates the loop currents to balance direct coupling currents [161, 162].

6.3.7 Feeding Technique

Various feed structures like T-shaped power divider [163], U-shaped feed [164], L-shaped feed [165], crossed feed [166], four-shaped feed [167], coupled feed [168], capacitive feed [169], and mixed feed [170] are used to gain enhancement and to solve the multi-path channel problems.

6.3.8 Ground Branches/Utilization

Ground branches like inverted/non-inverted L and T have specific effect on low/high frequencies, and are created between radiators to control mutual coupling [171]. MIMO with surface current suppression and inter-element decorrelation was designed on FR-4 dielectric substrate (thickness 1.56 mm) of size 40×50 mm^2 [172]. Inverted/non-inverted L and T ground branches can be combined with matching stubs [173, 174] to further enhance isolation. A MIMO antenna with FR-4 dielectric substrate (thickness 0.8 mm) of size 100×125 mm^2 was designed for impedance match [175]. Inverted/non-inverted L and T ground branches may offer wide-band operation and improved isolations. MIMO antenna with this concept was designed on FR-4 dielectric substrate (thickness 0.8 mm) of size 50 × 55 mm^2 in Fig. 6.3 [176].

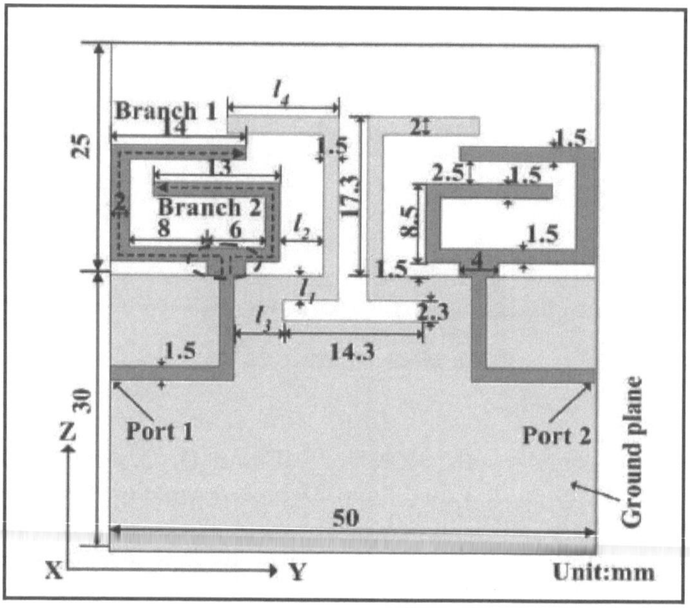

FIGURE 6.3
MIMO antenna with T-slot and L-branches [176]. [From: X. X. Xia, Q. X. Chu, and J. F. Li, "Design of a compact wideband MIMO antenna for mobile terminals," Progress in Electromagnetic Research C, vol. 41, pp. 163–174, July 2013. Reproduced courtesy of the Electromagnetics Academy.]

Multiple grounds in a MIMO and coils result in extra resonance to control the coupling among radiators [177]. Embedded protruded ground arm between the radiating elements helps in bandwidth and isolation improvement [178]. MIMO with inverted L-branches and a loop etched in a ground was designed on FR-4 dielectric substrate (thickness 0.8 mm) of size 85 × 125 mm^2, for isolation improvement [179]. Directive multi-loop MIMO utilizes the effect of reflector and controls surface currents. The port isolation may be greater than 15 dB here.

Different diversity and mutual coupling reduction techniques are compared with their diversity performance using frequency bands/resonant frequencies, number of elements, size, isolation, and ECC in Table 6.1. The amount of isolation is dependent on the type of radiating elements, diversity technique, placement of the radiators, their orientation, and the type of mutual coupling reduction technique used. The merits and demerits of diversity and mutual coupling reduction techniques are compared in Table 6.2.

TABLE 6.1
Some MIMO antennas with LTE, WLAN, and WiMAX bands

Ref. No.	Frequency/ bands (GHz)	No. of elements	Size (mm^2)	Isolation (dB)	ECC	Gain (dBi)	Isolation approach
[79]	2.02–2.93/ 5.10–6.45	2	100 × 50	>20	$<10^{-3}$	—	Space diversity
[83]	0.45	2	182 × 204	>20	$<10^{-2}$	—	Polarization and pattern diversity
[84]	2.65	3	120 × 90	>20	<0.4	2.8	Polarization diversity
[87]	2.4	2	50 × 50	—	—	—	Polarization diversity
[90]	2.4–4.8	4	114 × 114	>10	—	5 (Peak)	Polarization diversity
[92]	0.86	2	173 × 70	>40	$<10^{-2}$	>1.15	Polarization diversity
[94]	2.45	4	105 × 55	>10	$<10^{-2}$	2.15	Polarization diversity
[96]	1.8–2.9	4	140 × 120	>15	<0.1	>5.5	Slot etching
[102]	2.4–2.5/ 5.15–5.83	2	100 × 150	>15	0.1	—	Parasitic element
[112]	2.4–2.48/ 5.15–5.73	2	100 × 50	>25	—	—	Slot etching
[115]	2.4–2.5	4	80 × 60	>25	$<10^{-2}$	>2.84	Slit/slot etching
[120]	2.4	2	100 × 100	>40	—	>3.4	Slot etching
[149]	2.48	4	100 × 50	>10	<0.3	-0.8	CSRR
[152]	2.45/5.0	2	100 × 70	>18	$<10^{-2}$	—	Slit/slot etching
[158]	1.96–2.36	4	200 × 200	>24	—	>4	Shorting post
[161]	2.4/5.5	2	100 × 60	>26	$<10^{-2}$	4.84(Peak)	Shorting strips
[163]	2.4	2	60 × 90	>29	<0.2	—	Feed network
[166]	2.4	4	140 × 140	—	—	>1.8	Feed network
[168]	1.71–2.69	2	118 × 58	>12	<0.15	>1	Feed network
[170]	1.71–1.88	2	80 × 50	>15	$<10^{-2}$	>2.18	Feed network
[171]	1.88–2.20	4	95 × 60	>11.5	—	<1	L/T ground branches

6.4 MIMO Antenna Miniaturization Techniques

MIMO antennas with miniaturization result in strong mutual coupling. Therefore, different mutual coupling reduction techniques are used. Space utilization and compactness are the two issues associated with the MIMO antenna designs and require optimization for the coming generations. Due to miniaturization, transformation of antenna energy affects the desired performance parameters of the designed antenna. Different mutual coupling

TABLE 6.2
Merits and demerits of diversity and mutual coupling reduction techniques

No.	Technique's name	Merits	Demerits
1	Space diversity	–Easy –Reduced coupling	–Large size –Polarization loss
2	Polarization diversity	–Enhanced capacity –Compactness –Link improvement –High efficiency –Better isolation	–Polarization loss
3	Pattern diversity	–Easy –Link quality –Distinguished patterns	–De-couplers are required –Polarization impurity
4	Parasitic element	–Easy –Parasitic element –Limited ohmic losses	–Dependent on position/placement of parasitic elements
5	Neutralization line	–Easy –Creates shorting path of arms	–Very challenging placement
6	Slot etching	–Size reduction – Provides multi–banding –Broad-band –Isolation enhancement	–Back lobe
7	Decoupler	–Enhances isolation –Converts trans-admittance into imaginary part	–Placement dependent
8	SRR/CSRR	– Reduces size –Enhanced isolation –Band shifter	–Narrow band –Low efficiency –Low computational speed
9	Feed networks	–High isolation –High gain –High efficiency	–Large size

reduction techniques also provide the miniaturization of the overall size of MIMO antenna. These approaches are thoroughly discussed in this section. There are mainly six miniaturization approaches that may be used for space reduction in MIMO antenna designs. These are as follows:

1. *Dielectric material loading:* High degree of miniaturization is achieved here at the cost of required return-loss bandwidth. These materials are expensive and extra care is required to design and utilize them [180].

2. *Folding and shorting pins/posts:* This approach is used to make electrically small antennas for the wireless application. Meander line may be combined with PIFA for size reduction. This approach takes the benefit of reactive impedance that exists due to the short circuiting in PIFA antennas [181], and also in inverted F-antenna (IFA) with stub loading [182]. There is no design rule in this approach. Folding may reduce the four times of size of an antenna.

3. *Patch geometry and slots:* Different patch shapes provide the different factors of size reduction. Slots are also used for size reduction. Especially, fractal structures provide eight times of size reduction and wider bandwidths [183, 184]. Reconfigurability in slot and patch result in miniaturization of the designed MIMO [185, 186].

4. *Ground modification:* Ground modification technique is simple and it reduces eight times of size. No specific design rule is used here. Miniaturization is achieved at the cost of low efficiency and back radiation lobes [187–189].

5. *Metamaterials:* Metamaterial inspired shapes are used to induce ϵ_r negative (ENG), μ_r negative (MNG), and double negative (DNG) effects in the designed MIMO for size reduction. SRR and CSRR shapes are the well known examples of metamaterials [190, 191].

6. *Combination of asymmetric antennas:* Different shapes of patches are used to form the minaturized antenna. Such approach has the benefit of wider frequency response, and better diversity performance [192, 193].

6.5 Concluding Remarks

MIMO antennas have much better gain, efficiency, far-field characteristics, bandwidth, and multi-path solution in comparison with SISO. Radiation and emission properties are responsible for mutual coupling in compact MIMO antenna designs. Various techniques of mutual coupling reduction are used according to the requirement in designed MIMO. These techniques also provide the compactness of overall antenna design. Finally, the used mutual coupling reduction technique adds the effect of inductor or capacitor or forms the L–C equivalent circuit at the appropriate space for de-correlation and compactness [194].

7

Design and Analysis of Multi-Band Printed MIMO Antenna with Diversity and PSG

In this chapter, multi-band MIMO antenna with G-shaped folded monopole and partially stepped ground is presented and discussed to continuously satisfy the demand of high data rate and committed link for indoor and outdoor for WLAN/WiMAX applications. The G-shaped MIMO with 50.0 Ω ports is demonstrated and discussed with polarization diversity with the critical parametric evaluation and possible diversity effects. The combined effect of polarization diversity and folding enhances the compactness of designed MIMO. The simulated results are validated using VNA and in anechoic chamber to justify the design for real time applicability.

7.1 Introduction and Related Work

Fourth generation of systems particularizes the prerequisite of 1 Gbps for low-mobility users and 100 Mbps for high-mobility users. Various internet speeds are decided by ITU for small households, families with multiple devices, and an Internet devotee. The 76 Mbps speed is considered to be the best to administer multi-user downloading and streaming.

MIMO is the solution of multi-path propagation and is used for NLOS communication. Radio channel link is unpredictable in nature, due to this, link reliability, antenna gain stability, signal to noise ratio, and data rate becomes the function of NLOS. Loss of signaling in unpredicted channel leads to unnecessary delays, and creates break-down of reliable communication, due to very high fluctuations in signals used for wireless and mobile communication. Compactness of MIMO antennas is another issue in portable and mobile devices. Polarization diversity takes care of link quality, compactness, isolation between radiating elements, and adds a certain part in capacity improvement. Maintaining the proper isolation among closely packed elements and keeping them on the same substrate are very ambitious issues and obligations of MIMO antennas.

A 1×1 G-shaped monopole with coaxial feed on substrate of size 40×30 mm^2 was reported in the literature [195]. A 2×1 compact MIMO antenna with polarization diversity achieved more than 23 dB isolation on FR-4 substrate of size 70 × 35 mm^2 for wireless communication [196]. Variety of radiation patterns may be created with angular diversity by using reflectors and switches. Isolation in such cases is more than 6.7 dB between different ports, and substrate size is 100 × 120 mm^2 [197].

In this chapter a compact MIMO antenna with G-shaped folded monopole with polarization diversity is presented to cover IEEE 802.11 WLAN bands (2.4/5.2 GHz), and IEEE 802.16 WiMAX bands (2.5/5.5 GHz). The effect of partially stepped ground (PSG) is included to improve isolation and return-loss.

7.2 Multi-Band MIMO Antenna Design and Implementation

Four-port MIMO antenna with G-shaped elements and PSG is presented. Folded monopole radiating element reduces the overall size to increase the compactness. Low-cost FR-4 dielectric (substrate of permittivity 4.4, loss tangent of 0.025, and thickness of 1.524 mm) of size 70 × 70 mm^2 is utilized for the MIMO antenna design. All the radiating elements with 50.0 Ω ports and corresponding PSG grounds are orthogonally arranged to trans-receive signals in both vertical and horizontal directions.

The frequency of operation = 2.4/5.2/5.5 GHz, VSWR = 2:1, isolation >10 (dB), gain (dBi) >5, efficiency (%) >90, ECC <0.1, and WLAN/WiMAX applications are the set goals of the MIMO antenna design here. The optimized parameters are given in Table 7.1. A single G-shaped monopole with PSG ground structure is shown in Fig. 7.1 and fabricated 2 × 2 antenna is shown in Fig. 7.2.

TABLE 7.1
Optimized dimensions of presented MIMO (Unit: mm) [198].

Parameter	a	b	c	d	e	f	g
Value	28.40	8.58	12.62	5.92	18.93	15.39	2.96
Parameter	h	I	j	k	l	m	lx
Value	26.0	11.66	19.74	11.60	7.43	1.48	7.2

[From: Leeladhar Malviya, Rajib K. Panigrahi, and Machavaram V. Kartikeyan, "A 2 × 2 Dual-band MIMO antenna with polarization diversity for wireless applications," Progress in Electromagnetics Research C, vol. 61, pp. 91-103, January 2016. Reproduced courtesy of the Electromagnetics Academy.]

MIMO antenna resonates in lower-frequency band by virtue of the effect of the sum of lengths of arms $a + b + c + d + e + f$, where value of b = 7.10 mm. For higher-frequency band ground slot and extended arm b plays a very important role. Extension in arm b controls the return-loss around

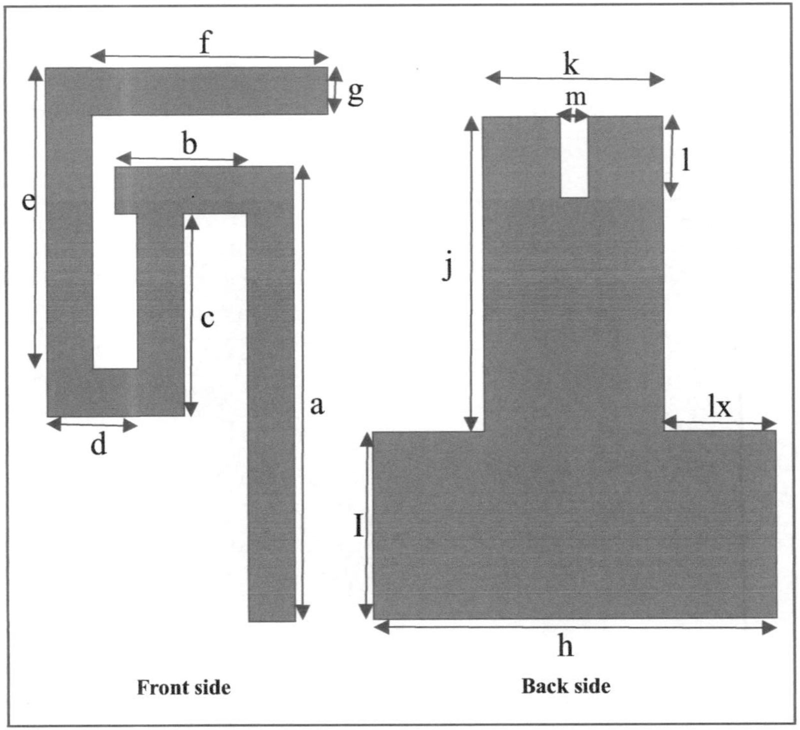

FIGURE 7.1
G-shaped radiating element and PSG ground [198]. [From: Leeladhar Malviya, Rajib K. Panigrahi, and Machavaram V. Kartikeyan, "A 2 × 2 Dual-band MIMO antenna with polarization diversity for wireless applications," Progress in Electromagnetics Research C, vol. 61, pp. 91–103, January 2016. Reproduced courtesy of the Electromagnetics Academy.]

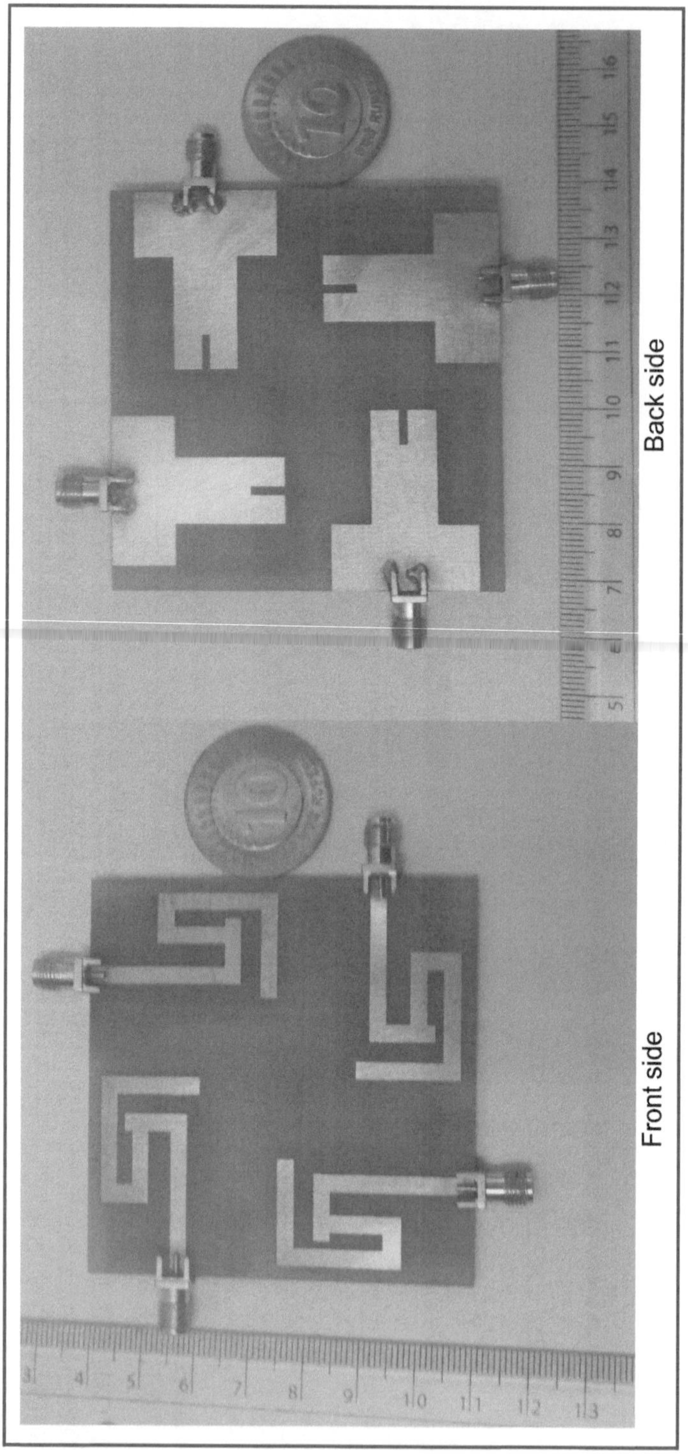

FIGURE 7.2
Fabricated 2×2 MIMO antenna [198]. [From: Leeladhar Malviya, Rajib K. Panigrahi, and Machavaram V. Kartikeyan, "A 2×2 Dual-band MIMO antenna with polarization diversity for wireless applications," Progress in Electromagnetics Research C, vol. 61, pp. 91–103, January 2016. Reproduced courtesy of the Electromagnetics Academy.]

5.0 GHz resonant frequency. Joint operation of ground slot and extended arm b results in lower return-loss. The length of extended arm b is 8.58 mm for the presented MIMO antenna.

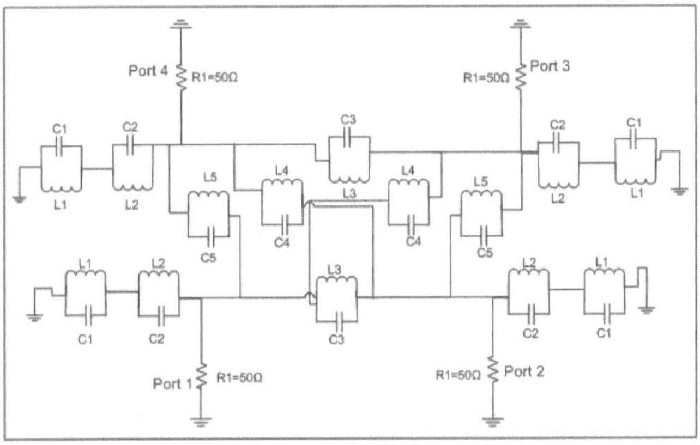

FIGURE 7.3
Equivalent circuit of presented MIMO [198]. [From: Leeladhar Malviya, Rajib K. Panigrahi, and Machavaram V. Kartikeyan, "A 2 × 2 Dual-band MIMO antenna with polarization diversity for wireless applications," Progress in Electromagnetics Research C, vol. 61, pp. 91–103, January 2016. Reproduced courtesy of the Electromagnetics Academy.]

The equivalent circuit parameters are extracted using the CST software to show the tuning behavior of the design. In general, a slot and gap are the capacitive bodies, while a narrow line is an inductive body. Therefore, at GHz frequencies slot is represented by parallel combination of the inductor (L) and capacitor (C), while resistor represents conduction and radiation losses. MIMO antenna port is represented by 50.0 Ω load. Parallel combination of components L1–C1 and L2–C2 represent the lower and higher resonant peaks (S_{11}). Mutual coupling effects (S_{12}, S_{13}, and S_{14}) are represented by parallel combination of L3–C3, L4–C4, and L5–C5, as shown in Fig. 7.3. The values of these circuit parameters are given as: L1 = 0.43 nH, C1 = 1.59 pf, L2 = 0.97 nH, C2 = 10.83 pf, L3 = 0.6 nH, C3 = 2.07 pf, L4 = 1.8 nH, C4 = 0.28 pf, L5 = 0.6 nH, and C5 = 2.07 pf.

7.3 Simulation-Measurement Results and Discussion

Particle swarm organization (PSO) is used to set minimum return-loss and low mutual coupling at MIMO ports. Vector network analyzer (VNA)-HP8720B measures return-loss and isolation of presented MIMO.

The return-loss parameter of a single element and 2×2 MIMO antenna are shown in Fig. 7.4 for the effectiveness of MIMO structure. A single radiating element without the PSG slot resonates in 2.35–2.60 GHz, and 4.71–4.97 GHz, in 2:1 impedance bands. Whereas, a single radiating element with PSG slot resonates in 2.35–2.62 GHz, and 4.87–5.56 GHz, in 2:1 impedance bands. Single radiating element with PSG slot covers 2.4/2.5/5.2/5.5 GHz WLAN/WiMAX frequency bands. Transformation of a single element in 2 × 2 MIMO antenna witness the reduced return-loss for 2.4/2.5/5.2/5.5 GHz WLAN/WiMAX frequency bands.

Due to four ports and orthogonal arrangement of MIMO antenna elements, CST-MWS and VNA exhibits $S_{11} = S_{22} = S_{33} = S_{44}$, $S_{12} = S_{21} = S_{34} = S_{43} = S_{14} = S_{41} = S_{23} = S_{32}$, and $S_{13} = S_{31} = S_{24} = S_{42}$. For easy analysis of diversity parameters, consider S_{11}, S_{12}, and S_{13} scattering parameters only, throughout the chapter.

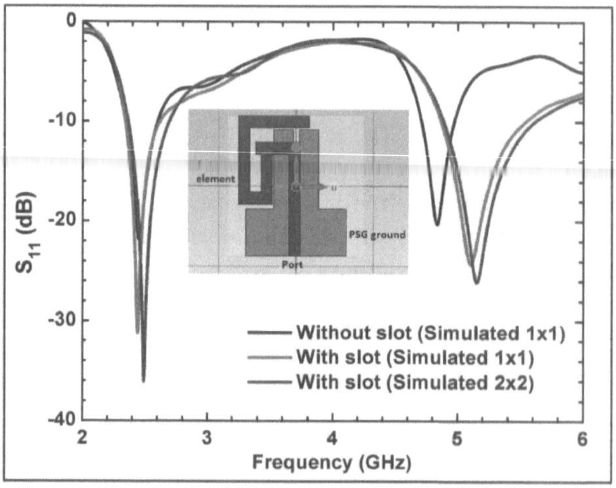

FIGURE 7.4

S_{11} parameter of single element and 2×2 MIMO antenna [198]. [From: Leeladhar Malviya, Rajib K. Panigrahi, and Machavaram V. Kartikeyan, "A 2 × 2 Dual-band MIMO antenna with polarization diversity for wireless applications," Progress in Electromagnetics Research C, vol. 61, pp. 91–103, January 2016. Reproduced courtesy of the Electromagnetics Academy.]

Figure 7.5 shows the simulated and measured S parameters of dual-band MIMO antenna. The simulated 2:1 VSWR bands extend from 2.37–2.69 GHz and 4.89–5.61 GHz frequencies, and measured 2:1 VSWR bands extend from 2.41–2.78 GHz, and 4.96–5.64 GHz frequencies. The simulated and measured isolations at different ports in lower and higher-frequency bands are more than 17 dB. The simulated bandwidth in lower and higher bands are 320 MHz and 720 MHz, whereas measured bandwidth in lower and higher bands are 368

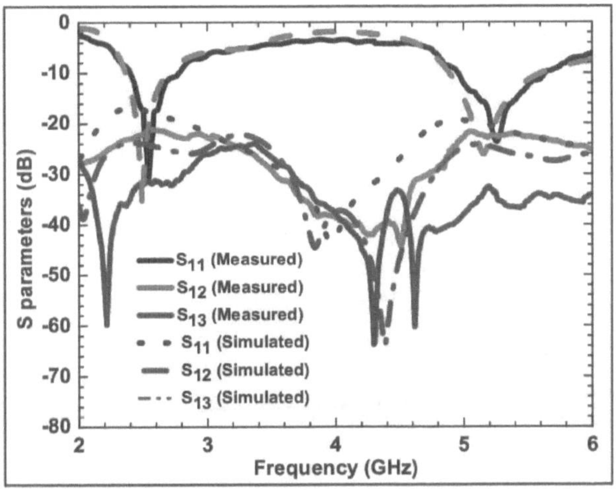

FIGURE 7.5
S parameters of 2 × 2 MIMO antenna [198]. [From: Leeladhar Malviya, Rajib K. Panigrahi, and Machavaram V. Kartikeyan, "A 2 × 2 Dual-band MIMO antenna with polarization diversity for wireless applications," Progress in Electromagnetics Research C, vol. 61, pp. 91–103, January 2016. Reproduced courtesy of the Electromagnetics Academy.]

MHz and 680 MHz. The difference in simulated and measured results are due to the fabrication errors, and port/cable coupling losses.

Amount of isolation depends on the linking of surface current to the un-excited ports (all un-excited ports are terminated by 50.0 Ω). From Fig. 7.6, it is observed that high current is linked on PSG arm and closest antenna element arm, and less current is linked with the un-excited radiating elements. Such effect is exhibited by S_{12}, S_{13} and S_{14} (where $S_{12} = S_{14}$) isolation parameters (ports 2 and 4 are equi-distant from excited port 1, and port 3 is away from it). This effect is counted in the lower-frequency band, where isolation is better than 17 dB, and simulated surface current distribution (SCD) is in the range of 0–110 A/m at 2.54 GHz resonant frequency. The same concept is followed when any other port or all the ports are excited.

Figure 7.7 exhibits most of the current on the excited element (port 1), and some are coupled to the un-excited port elements. This is counted as a higher-frequency band, where isolation is more than 21 dB at different ports. The simulated SCD at 5.26 GHz resonant frequency is in the range of 0–44.5 Ampere/metre. During the excitation of other ports, the same concept is followed.

Figure 7.8 shows the comparison of *S* parameters of full and PSG grounds. MIMO with full ground resonates at 3.8 GHz frequency, where, impedance band is 3.80–3.91 GHz (bandwidth is 112 MHz). Isolation parameters S_{12} and S_{13} have high values with the full ground. MIMO with PSG achieves

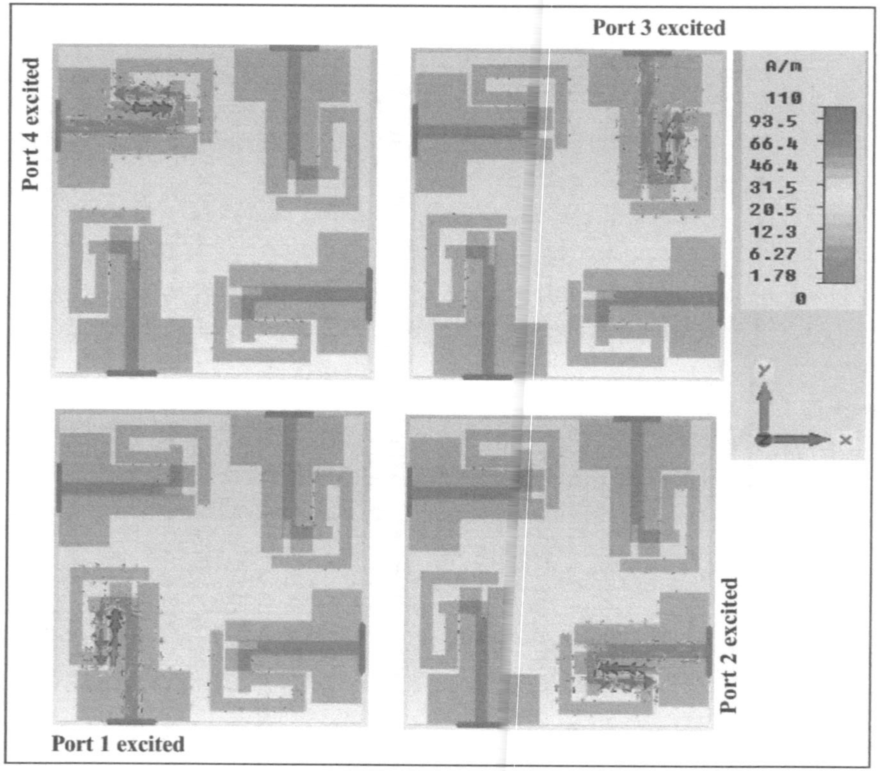

FIGURE 7.6
SCD of MIMO antenna at 2.54 GHz [198]. [From: Leeladhar Malviya, Rajib K. Panigrahi, and Machavaram V. Kartikeyan, "A 2 × 2 Dual-band MIMO antenna with polarization diversity for wireless applications," Progress in Electromagnetics Research C, vol. 61, pp. 91–103, January 2016. Reproduced courtesy of the Electromagnetics Academy.]

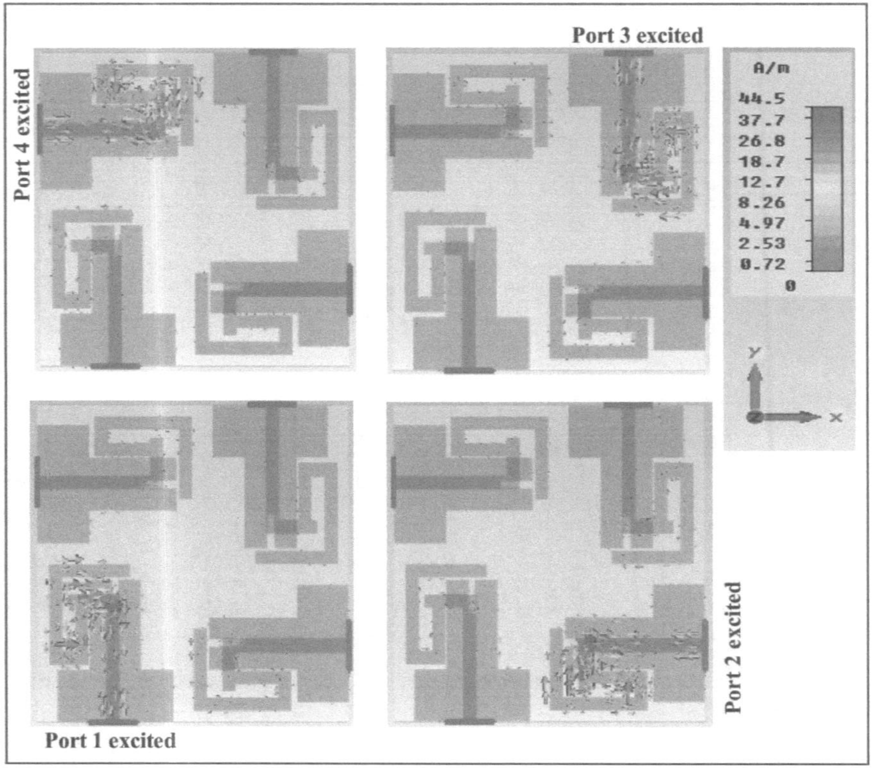

FIGURE 7.7
SCD of MIMO antenna at 5.26 GHz [198]. [From: Leeladhar Malviya, Rajib K. Panigrahi, and Machavaram V. Kartikeyan, "A 2 × 2 Dual-band MIMO antenna with polarization diversity for wireless applications," Progress in Electromagnetics Research C, vol. 61, pp. 91–103, January 2016. Reproduced courtesy of the Electromagnetics Academy.]

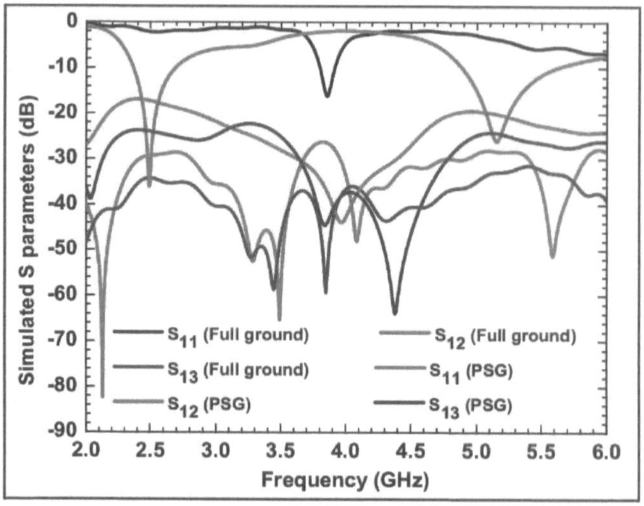

FIGURE 7.8
S parameters of full ground and PSG ground [198]. [From: Leeladhar Malviya, Rajib K. Panigrahi, and Machavaram V. Kartikeyan, "A 2 × 2 Dual-band MIMO antenna with polarization diversity for wireless applications," Progress in Electromagnetics Research C, vol. 61, pp. 91–103, January 2016. Reproduced courtesy of the Electromagnetics Academy.]

compactness and desired resonant characteristics with adequate isolations among ports.

Figure 7.9 exhibits the effect of ground slot length variations on S parameters (slot width is fixed at 1.48 mm). From the parametric evaluation of ground slot length (l), it is observed that MIMO with no slot in ground, resonates at 2.5 and 4.8 GHz frequencies. After the parametric analysis, a slot is etched in ground to get resonant around 5.0 GHz frequency. The length of the slot is selected as 7.43 mm from the top of the PSG ground. Except for S_{11}, no big changes are seen in S_{12}, and S_{13} isolation parameters.

Figure 7.10 exhibits S parameters with arm b evaluation. To achieve minimum return-loss around 5.0 GHz, the length of arm b is fixed at 8.58 mm. For the value of b = 8.21 mm, very low return-loss is observed in the first band, and a small peak at 4.35 GHz. Certain variations in S_{11} are observed in different bands with evaluation of arm length b, and minute changes in S_{12}, and S_{13} isolation parameters.

Figure 7.11 exhibits simulated responses of the equivalent circuit using ADS and CST simulated responses. CST and circuit simulated responses have minute differences in lower- and higher-frequency bands for S_{11}, S_{12}, and S_{13} parameters.

2×2 MIMO antenna diversity parameters are MIMO gain, efficiency, ECC, TARC, and MEG. Measurement of MIMO antenna gain is carried out in an

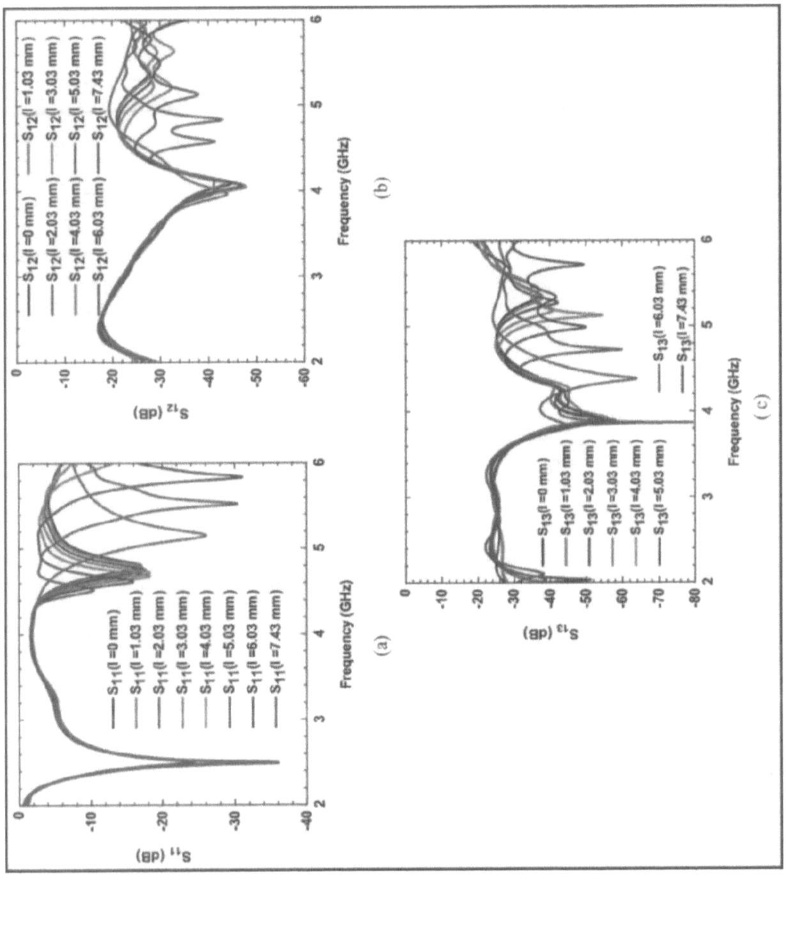

FIGURE 7.9
S parameters with ground slot length [198]. [From: Leeladhar Malviya, Rajib K. Panigrahi, and Machavaram V. Kartikeyan, "A 2×2 Dual-band MIMO antenna with polarization diversity for wireless applications," Progress in Electromagnetics Research C, vol. 61, pp. 91–103, January 2016. Reproduced courtesy of the Electromagnetics Academy.]

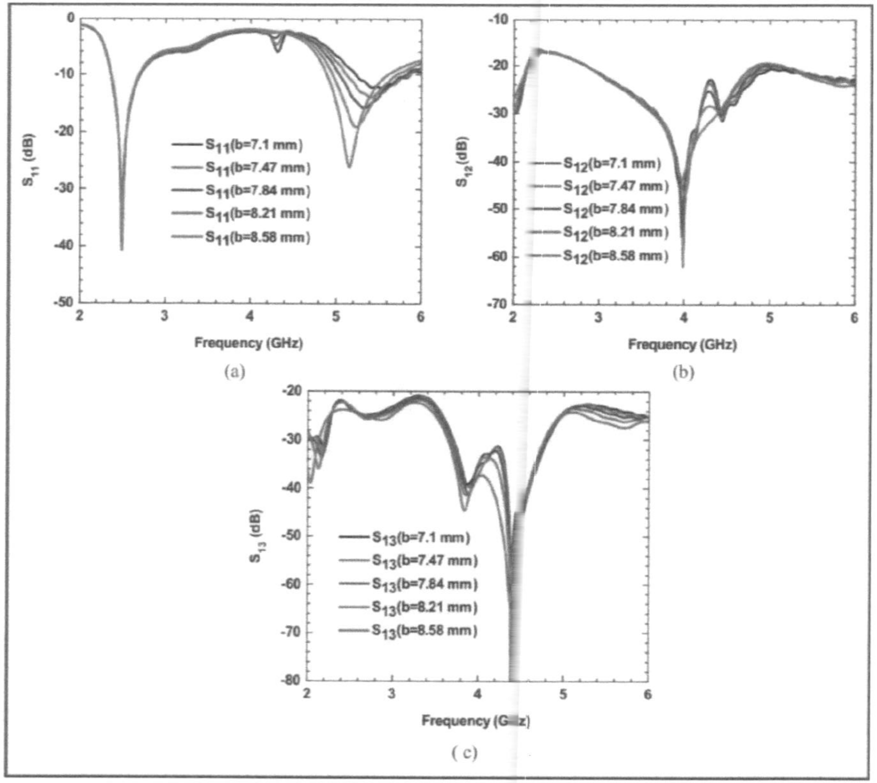

FIGURE 7.10
S parameters with arm b [198]. [From: Leeladhar Malviya, Rajib K. Panigrahi, and Machavaram V. Kartikeyan, "A 2×2 Dual-band MIMO antenna with polarization diversity for wireless applications," Progress in Electromagnetics Research C, vol. 61, pp. 91–103, January 2016. Reproduced courtesy of the Electromagnetics Academy.]

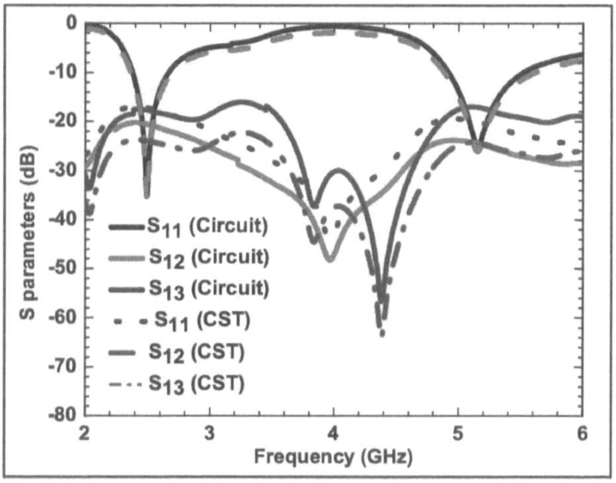

FIGURE 7.11
Comparison of circuit and CST S parameters [198]. [From: Leeladhar Malviya, Rajib K. Panigrahi, and Machavaram V. Kartikeyan, "A 2 × 2 Dual-band MIMO antenna with polarization diversity for wireless applications," Progress in Electromagnetics Research C, vol. 61, pp. 91–103, January 2016. Reproduced courtesy of the Electromagnetics Academy.]

anechoic chamber using the substitution method (with two standard horn antennas and designed antenna). The single radiating element shows better than 2.0 dBi simulated gain in both the frequency bands. Presented MIMO antenna shows simulated gains of 3.95 dBi and 4.02 dBi at 2.54 GHz and 5.26 GHz resonant frequencies. Measured gains at 2.54 GHz and 5.26 GHz resonant frequencies are 3.98 dBi and 4.13 dBi. Measured and simulated MIMO antenna gains are shown in Figs. 7.12 and 7.13.

For single radiating element simulated radiation efficiency is more than 85% in the lower band, and more than 73.49% in the higher band. Gain-efficiency plot of the presented MIMO shows 88.42% and 74.94% radiation efficiencies at 2.54 GHz and 5.26 GHz resonant frequencies. Lower and higher bands have more than 68% radiation efficiency. Total efficiency in lower- and higher-frequency bands is more than 60%.

ECC is obtained using (4.7). The values of $i = 1$ to 2, $j = 1$ to 2 for considered two elements, and $N = 2$ (selected group of antennas). ECCs at different ports are denoted by ρ_{e12}, ρ_{e13}, and ρ_{e14}. Where, ρ_{e12} is the ECC between ports 1–2, ρ_{e13} is between ports 1–3, and ρ_{e14} is between ports 1–4. ECCs for the presented design are obtained by substituting the $S_{11} = S_{22} = S_{33} = S_{44}$, $S_{12} = S_{21} = S_{34} = S_{43} = S_{14} = S_{41} = S_{23} = S_{32}$, and $S_{13} = S_{31} = S_{24} = S_{42}$ in (4.7).

Figures 7.14 and 7.15 show the simulated ECC, which lies in the range 0–0.01, for lower and higher bands. Here, measured ECC between any of two

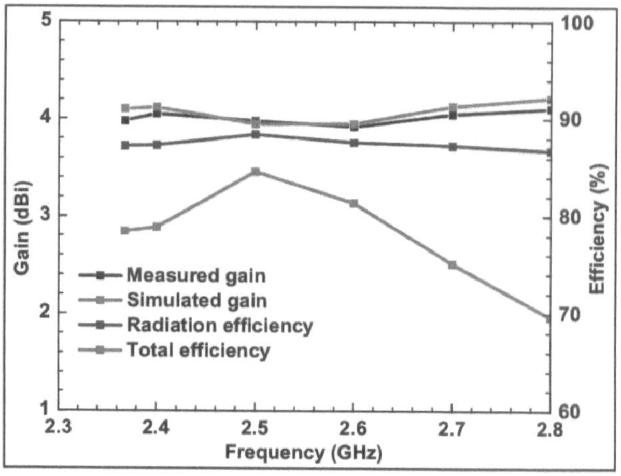

FIGURE 7.12
Gain and efficiency in lower band [198]. [From: Leeladhar Malviya, Rajib K. Panigrahi, and Machavaram V. Kartikeyan, "A 2 × 2 Dual-band MIMO antenna with polarization diversity for wireless applications," Progress in Electromagnetics Research C, vol. 61, pp. 91–103, January 2016. Reproduced courtesy of the Electromagnetics Academy.]

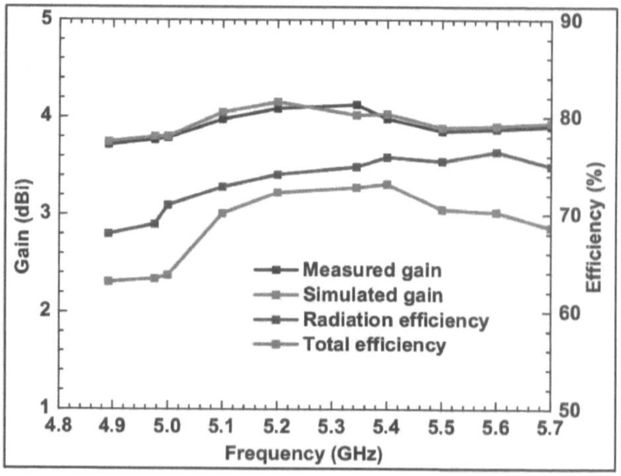

FIGURE 7.13
Gain and efficiency in higher band [198]. [From: Leeladhar Malviya, Rajib K. Panigrahi, and Machavaram V. Kartikeyan, "A 2 × 2 Dual-band MIMO antenna with polarization diversity for wireless applications," Progress in Electromagnetics Research C, vol. 61, pp. 91–103, January 2016. Reproduced courtesy of the Electromagnetics Academy.]

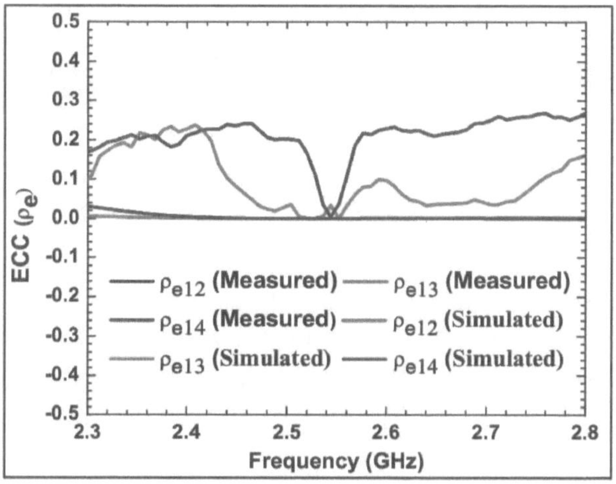

FIGURE 7.14
ECC in the lower band [198]. [From: Leeladhar Malviya, Rajib K. Panigrahi, and Machavaram V. Kartikeyan, "A 2 × 2 Dual-band MIMO antenna with polarization diversity for wireless applications," Progress in Electromagnetics Research C, vol. 61, pp. 91–103, January 2016. Reproduced courtesy of the Electromagnetics Academy.]

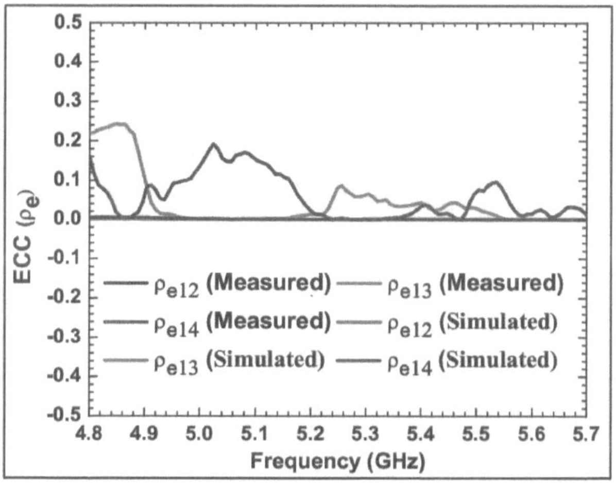

FIGURE 7.15
ECC in the higher band [198]. [From: Leeladhar Malviya, Rajib K. Panigrahi, and Machavaram V. Kartikeyan, "A 2 × 2 Dual-band MIMO antenna with polarization diversity for wireless applications," Progress in Electromagnetics Research C, vol. 61, pp. 91–103, January 2016. Reproduced courtesy of the Electromagnetics Academy.]

ports lies in the range 0–0.27 for both bands. At 2.54 GHz resonant frequency, the simulated ECC values are $\rho_{e12} = 2.54\times10^{-4}$, $\rho_{e13} = 7.54\times10^{-6}$, and $\rho_{e14} = 2.54 \times 10^{-4}$. Similarly, the measured ECC values at 2.54 GHz frequency are $\rho_{e12} = 4.34\times10^{-3}$, $\rho_{e13} = 3.4\times10^{-2}$, and $\rho_{e14} = 4.34\times10^{-3}$. The values of ECC for 5.26 GHz resonant frequency for simulated results are $\rho_{e12} = 4.24 \times 10^{-4}$, $\rho_{e13} = 3.72 \times 10^{-6}$, and $\rho_{e14} = 4.24 \times 10^{-4}$. Similarly, the measured ECC values at 5.26 GHz frequency are $\rho_{e12} = 1.67 \times 10^{-3}$, $\rho_{e13} = 7.4 \times 10^{-2}$, and $\rho_{e14} = 1.67 \times 10^{-3}$.

Very less ECCs between any of the two radiators at resonant frequencies are achieved (measured results have some fluctuations). ECCs at ports 1-2 and 1-4 have equal values due to equi-distance from port 1. Therefore, only two curves of measured/simulated results are observable here. The difference in measured and simulated ECCs is due to the fabrication errors and port/cable coupling losses. ECC in both the frequency bands is lower than 0.3, which is also the requirement for 4G applications.

Similarly, TARC includes the random signals and phase angles at different ports by including interference and all the other disturbances due to Gaussian nature of signals. TARC can be obtained for pair of ports using (4.10). Figure 7.16 exhibits TARC with pair of excitation angles at different ports. Presented MIMO antenna with 90°, 180° of excitation signals, shows less distortion in the measured results. The shape of the TARC curves shows the resemblance with the S_{11} parameter. TARC bandwidths in lower and higher bands are approximately 300 MHz and 700 MHz for 2:1 VSWR.

The MEG is another diversity parameter and is measured using far-field analysis. MEG includes radiation efficiency and power patterns of the designed MIMO antenna. MEG values are computed using (4.15).

Let us consider horizontal and vertical components of Gaussian signals with mean $(\mu) = 0$ and variance $(\sigma) = 20$ respectively. Figure 7.17 shows the comparison between Isotropic and Gaussian mediums, for different values of $XPRs$ and MEGs using CST simulations. Due to indoor and outdoor environments, changes are observable for $XPR = 0$ dB and $XPR = 6$ dB. For different values of θ and ϕ, variation in received power is observable. The value of MEGs for 2.54 GHz and 5.26 GHz resonant frequencies for Isotropic medium for $XPR = 0$ dB and $XPR = 6$ dB are less than −4.68 dB, and for Gaussian medium, these values are less than −5.3 dB (due to the symmetry of radiating structures, value is common for each element). In practice, different values of MEGs may be achieved for different antenna elements. From the observations, presented MIMO antenna has good MEG characteristics for both the indoor and outdoor applications.

Far-field radiation pattern measurement is performed in an anechoic chamber in the presence of a standard horn antenna (transmitting antenna). During the measurement, any one of the radiating element is selected in receiving mode and remaining ports are terminated by 50.0 Ωs to avoid any signal pick up. Far-field radiation patterns of polarization diversity MIMO antenna are measured at resonant frequencies.

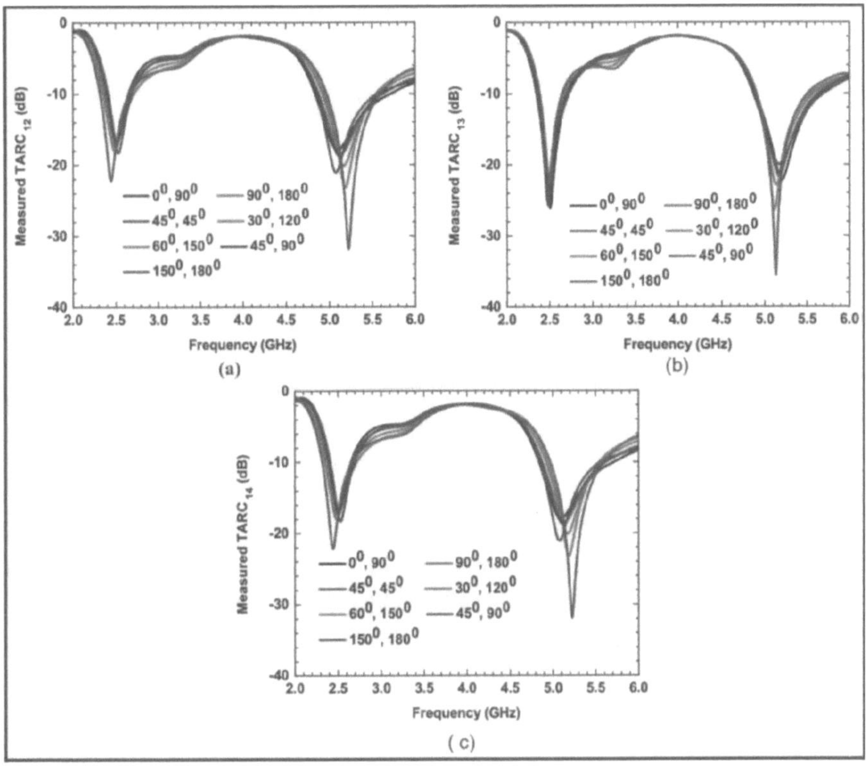

FIGURE 7.16
TARC of MIMO antenna [198]. [From: Leeladhar Malviya, Rajib K. Panigrahi, and Machavaram V. Kartikeyan, "A 2 × 2 Dual-band MIMO antenna with polarization diversity for wireless applications," Progress in Electromagnetics Research C, vol. 61, pp. 91–103, January 2016. Reproduced courtesy of the Electromagnetics Academy.]

Figures 7.18 and 7.19 exhibit simulated E-field and H-field patterns of single element at lower (2.44 GHz) and higher (5.104 GHz) resonant frequencies, and effectiveness of it. For 2 × 2 MIMO, E_θ, E_ϕ, H_θ, and H_ϕ components are considered. Effect of E-field and H-field on both the components (θ and ϕ) can easily be noticed. Figures 7.20 and 7.21 show the verification of polarization diversity at each ports with θ and ϕ components.

Figures 7.22 and 7.23 show the comparison between simulated and measured E-field and H-field patterns of presented MIMO antenna. Radiation patterns at ports 1–3 and 2–4 are complementary symmetric for each resonant case. Placement of one component in horizontal and other in vertical leads to the situation. Certain discrepancies between simulated and measured patterns is obvious due to the errors in fabrication process, and port/cable coupling losses.

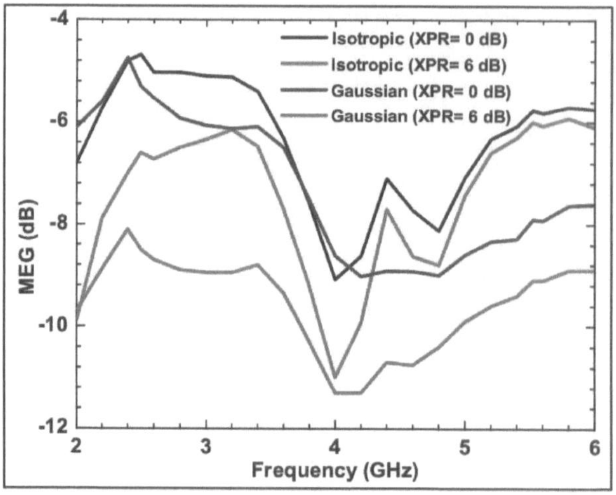

FIGURE 7.17
MEG comparisons for Isotropic and Gaussian mediums [198]. [From: Leeladhar Malviya, Rajib K. Panigrahi, and Machavaram V. Kartikeyan, "A 2 × 2 Dual band MIMO antenna with polarization diversity for wireless applications," Progress in Electromagnetics Research C, vol. 61, pp. 91–103, January 2016. Reproduced courtesy of the Electromagnetics Academy.]

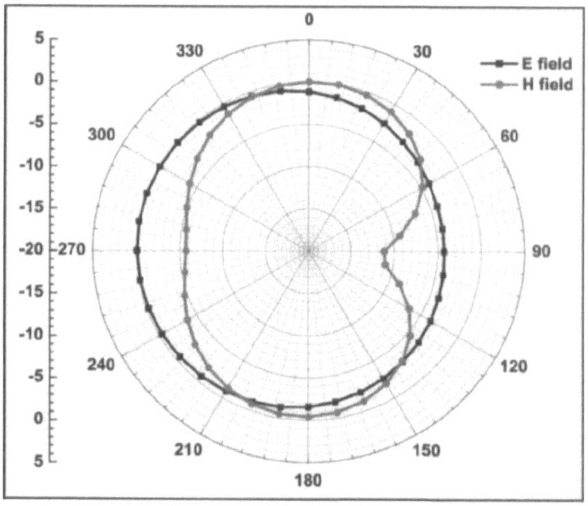

FIGURE 7.18
E-field and H-field patterns at lower resonant of single element [198]. [From: Leeladhar Malviya, Rajib K. Panigrahi, and Machavaram V. Kartikeyan, "A 2 × 2 Dual-band MIMO antenna with polarization diversity for wireless applications," Progress in Electromagnetics Research C, vol. 61, pp. 91–103, January 2016. Reproduced courtesy of the Electromagnetics Academy.]

Table 7.2 compares earlier reported works with designed MIMO. The design given in reference [79] has wide-band, low ECC, but belongs to large size for two elements only. The other designs given in table have, either large sizes or high ECCs in comparison with the designed MIMO.

TABLE 7.2
Comparison of designed MIMO with earlier reported works [198] [From: Leeladhar Malviya, Rajib K. Panigrahi, and Machavaram V. Kartikeyan, "A 2 × 2 Dual-band MIMO antenna with polarization diversity for wireless applications," Progress in Electromagnetics Research C, vol. 61, pp. 91–103, January 2016. Reproduced courtesy of the Electromagnetics Academy.]

Ref. No.	Frequency/ bands (GHz)	No. of elements	Size (mm^2)	Isolation (dB)	ECC	Gain (dBi)
[57]	2.45/5.8	2	100 × 50	>20	0.3	—
[79]	2.02–2.93 5.10–6.45	2	100 × 50	>20	$<10^{-3}$	—
[90]	2.4–4.8	4	114 × 114	>10	—	5(Peak)
[102]	2.4–2.5 5.15–5.83	2	100 × 150	>15	0.1	—
[112]	2.4–2.48 5.15–5.73	2	100 × 50	>25	—	—
[115]	2.4–2.5	4	80 × 60	>25	$<10^{-2}$	>2.84
[149]	2.48	4	100 × 50	>10	<0.3	−0.8
[152]	2.45/5.0	2	100 × 70	>18	$<10^{-2}$	—
[161]	2.4/5.5	2	100 × 60	>26	$<10^{-2}$	4.84(Peak)
[166]	2.4	4	140 × 140	—	—	>1.8
Designed MIMO	2.37–2.69 (Lower) 4.89–5.61 (Higher)	4	70 × 70	>17 >19	$<10^{-2}$	>3.6 >3.6

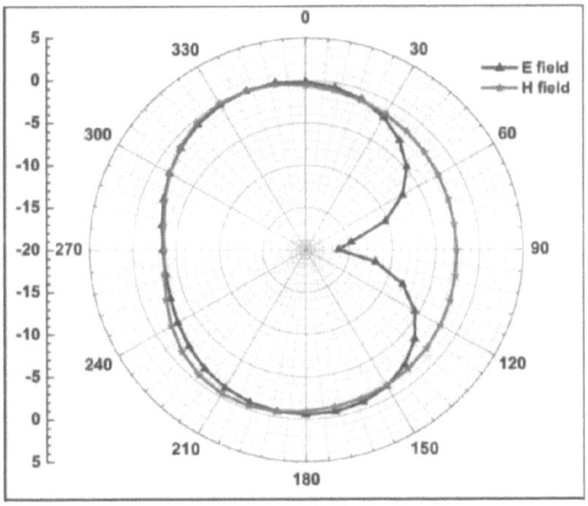

FIGURE 7.19
E-field and H-field patterns at higher resonant of single element [198]. [From: Leeladhar Malviya, Rajib K. Panigrahi, and Machavaram V. Kartikeyan, "A 2 × 2 Dual-band MIMO antenna with polarization diversity for wireless applications," Progress in Electromagnetics Research C, vol. 61, pp. 91–103, January 2016. Reproduced courtesy of the Electromagnetics Academy.]

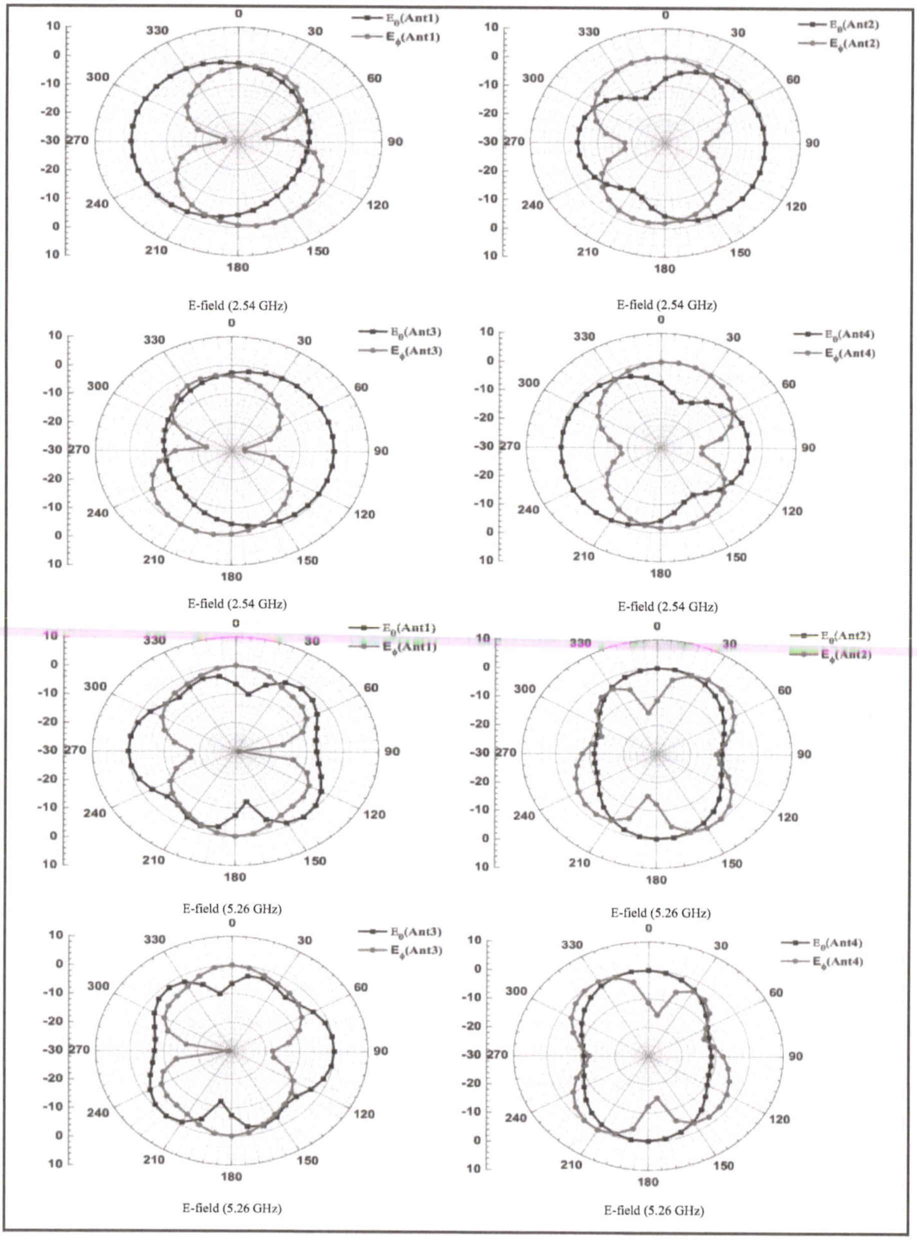

FIGURE 7.20
E_θ and E_ϕ radiation patterns of MIMO antenna [198]. [From: Leeladhar Malviya, Rajib K. Panigrahi, and Machavaram V. Kartikeyan, "A 2×2 Dual-band MIMO antenna with polarization diversity for wireless applications," Progress in Electromagnetics Research C, vol. 61, pp. 91–103, January 2016. Reproduced courtesy of the Electromagnetics Academy.]

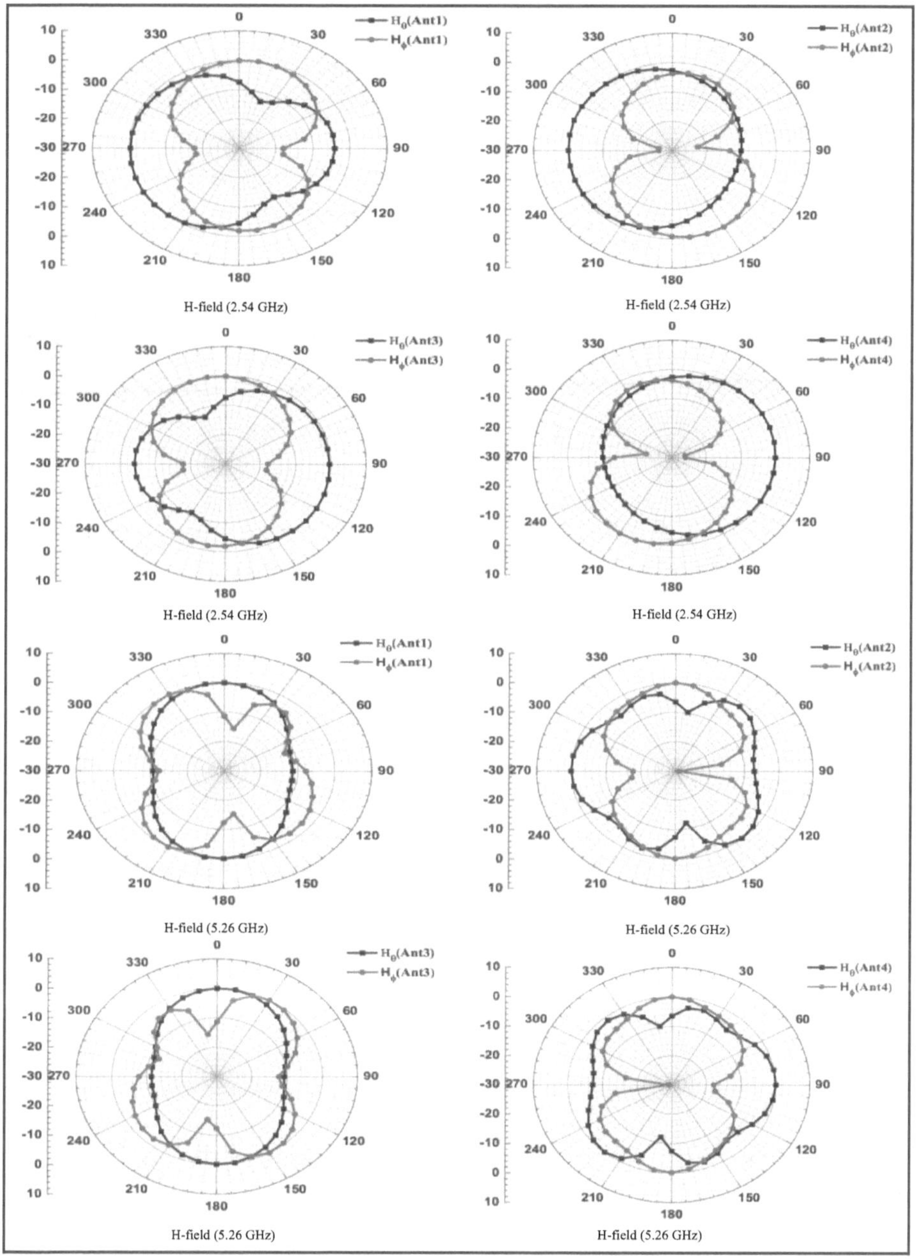

FIGURE 7.21
H_θ and H_ϕ radiation patterns of MIMO antenna [198]. [From: Leeladhar Malviya, Rajib K. Panigrahi, and Machavaram V. Kartikeyan, "A 2×2 Dual-band MIMO antenna with polarization diversity for wireless applications," Progress in Electromagnetics Research C, vol. 61, pp. 91–103, January 2016. Reproduced courtesy of the Electromagnetics Academy.]

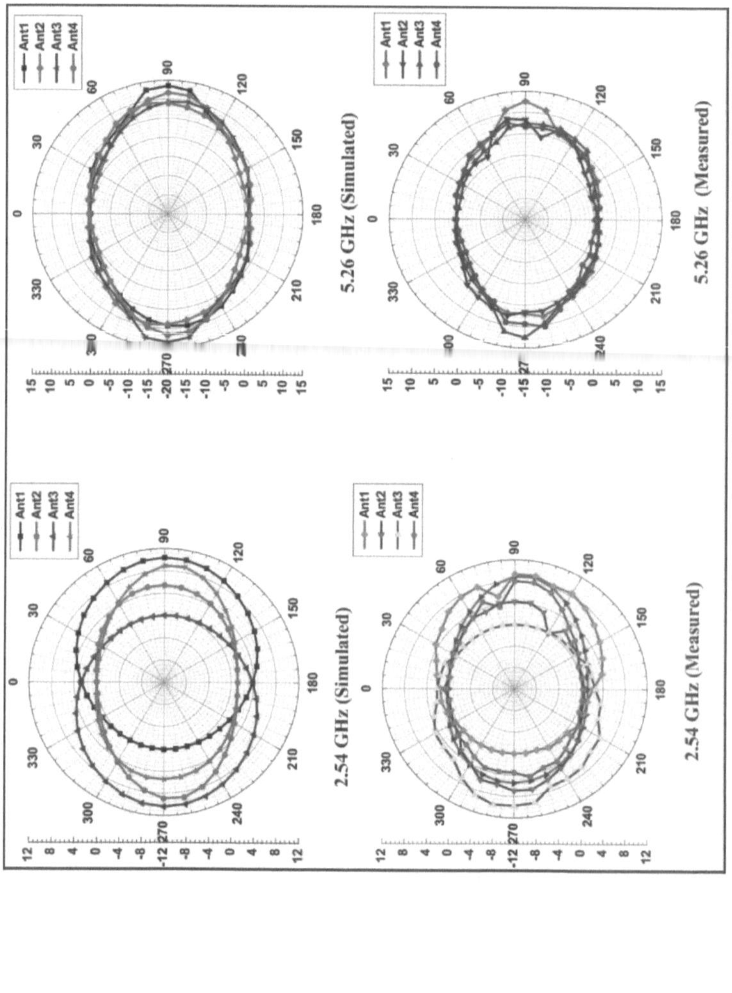

FIGURE 7.22
E-field radiation patterns of MIMO antenna [198]. [From: Leeladhar Malviya, Rajib K. Panigrahi, and Machavaram V. Kartikeyan, "A 2 × 2 Dual-band MIMO antenna with polarization diversity for wireless applications," Progress in Electromagnetics Research C, vol. 61, pp. 91–103, January 2016. Reproduced courtesy of the Electromagnetics Academy.]

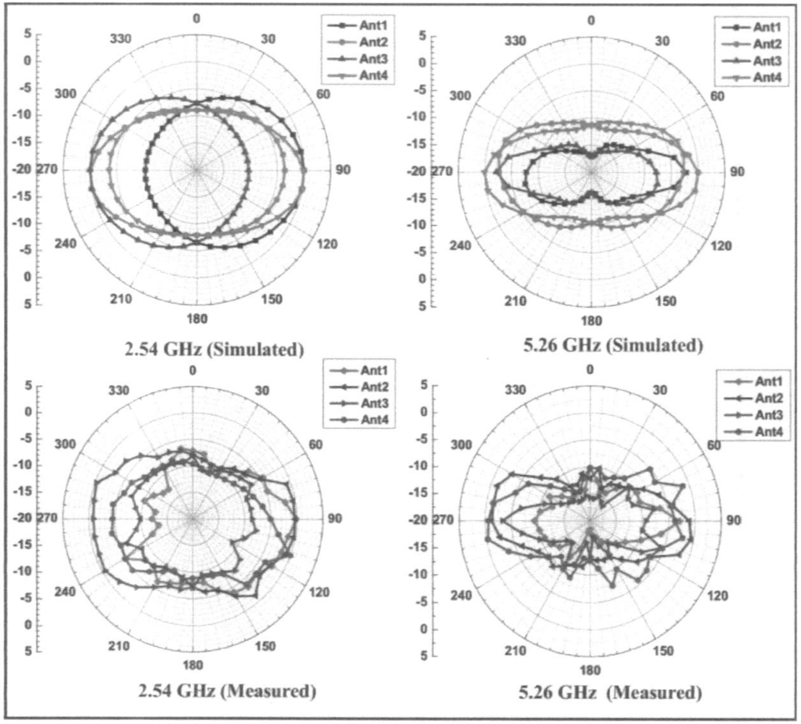

FIGURE 7.23
H-field radiation patterns of MIMO antenna [198]. [From: Leeladhar Malviya, Rajib K. Panigrahi, and Machavaram V. Kartikeyan, "A 2 × 2 Dual-band MIMO antenna with polarization diversity for wireless applications," Progress in Electromagnetics Research C, vol. 61, pp. 91–103, January 2016. Reproduced courtesy of the Electromagnetics Academy.]

7.4 Concluding Remarks

Four port MIMO antenna with polarization diversity technique and with G-shaped radiating elements and PSG has been presented for dual-band wireless applications. Jointly, polarization diversity and PSG ground control the effect of mutual coupling between radiating elements in 2.41–2.78 GHz and 4.96–5.64 GHz frequency bands. Measured isolation in lower and higher bands is more than 21 dB. Measured gains at 2.54 GHz and 5.26 GHz resonant frequencies are 3.98 dBi and 4.13 dBi, respectively. The designed MIMO antenna covers WLAN bands (2.4/5.2 GHz) and WiMAX bands (2.5/5.5 GHz) for wireless applications. The measured value of ECC, TARC, and MEG shows the good diversity performance. The presented design is applicable for both the indoor and outdoor environments [198].

8

Design and Analysis of Wide-Band MIMO Antenna with Diversity and PEG

Compact and diversified antennas are required to solve the difficulties of transmit and receive operations. Variety of solutions are practiced to control the degrading factors of mutual coupling. However, reliability is also one of the issues in wireless communication. For the coverage of the intended area, distinct field patterns are accomplished by pattern diversity. High throughput requirement in non-line-of-sight (NLOS) communication is satisfied by the collocated multiple-input multiple-output (MIMO) radiators. However, increasing the radiating elements in MIMO places the upper limit on the achievable performance parameters. A wide-band MIMO antenna with diversity and PEG is presented in this chapter to cover all WLAN/WiMAX bands.

8.1 Introduction and Related Work

Variety of solutions are practiced to implement pattern diversity in multi-element antenna systems. Open-loop radiators, curved slots, and certain alphabetic structures can implement the pattern diversity effects. District shape antenna radiators in MIMO may create asymmetry, which can be exploited for compactness and absolute size of MIMO antenna [199–201]. MIMO was designed for universal mobile telecommunications (UMTs) application using asymmetric patch shapes. The combined effort of diversity techniques limits the adverse effects of mutual coupling and provides better than 11.5 dB of isolation among radiating elements [202, 203]. Mutual coupling effect in four-port MIMO can be limited using array structures (or combining them with arrays) [204]. A U-shaped slot and metamaterial inspired MIMO antennas may produce better isolation between radiating element [205].

In this chapter, four-port wide-band MIMO antenna with diversity and partially extended ground (PEG) is designed for wireless applications. Designed MIMO achieves better than 12.6 dB of isolation among radiating elements. The designed MIMO covers IEEE 802.11 a/h/j/n/ac/p/y WLAN standards including ISM band, and WiMAX bands (2.5/3.5/5.5 GHz) in this range.

8.2 Wide-Band MIMO Antenna Design and Implementation

Four-port MIMO antenna elements are implemented in such a manner that signals can be received in distinct directions. MIMO antenna with pattern diversity and wide-band characteristics is designed using CST simulator and FR-4 dielectric substrate (thickness of 1.524 mm, permittivity of 4.4, and loss tangent of 0.025) of size 54.82 × 96.09 mm^2 is used for fabrication. A combination of triangular and trapezoidal patches separated by inclined slot of 45° is used to design the MIMO antenna elements. Effect of 45° inclined slot can be accounted on bandwidth extension of the designed antenna [206], and to get distinct field patterns [207].

The designed ground structure in this chapter is explained in two steps. In first step, fractional ground is utilized to reduce the size of MIMO structure. Second part consists of three PEGs for isolation enhancement between radiators. The width of each PEG arm is 1.70 mm and the separation between the PEG arms is 0.54 mm. The slot separating triangular and trapezoidal patches has a width of 0.74 mm, and is created at a corner length of 1.28 mm (from the bottom side of the patch). This inclined slot adds the capacitive effect to enhance frequency response, and, also contributes in complementary field patterns for the coverage of WLAN/WiMAX frequency bands.

The frequency of operation = 2.0–6.0 GHz, VSWR = 2:1, isolation >10 dB, gain (dBi) >3, efficiency (%) >90, ECC <0.1, and WLAN/WiMAX applications are the set goals of the presented MIMO antenna design here. The optimized design parameters are given in Table 8.1. The schematic and fabricated views are shown in Fig. 8.1 and Fig. 8.2.

TABLE 8.1

Optimized parameters of wide-band MIMO (Unit: mm) [208] [From: Leeladhar Malviya, R. K. Panigrahi, and M. V. Kartikeyan, "A multi-standard, wide-band 2 × 2 compact MIMO antenna with ground modification techniques," International Journal of Microwave and Optical Technology (IJMOT), vol. 11, no. 4, pp. 259–267, July 2016. Reproduced courtesy of the International Journal of Microwave and Optical Technology.]

Parameter	a	b	c	d	e	f
Value	20.83	10.42	21.93	8.4	54.82	0.74
Parameter	g	h	i	j	k	l
Value	2.96	96.09	18.68	23.75	1.70	0.54

Equivalent circuit analysis is accomplished using ADS software. The tuning characteristics are the effect of patch, transmission line, and PEG ground. Each port is represented by a 50.0 Ω load. The patch/resonant peak (S_{11}) is the parallel combination of L1-C1 components. Isolation parameter S_{12} is

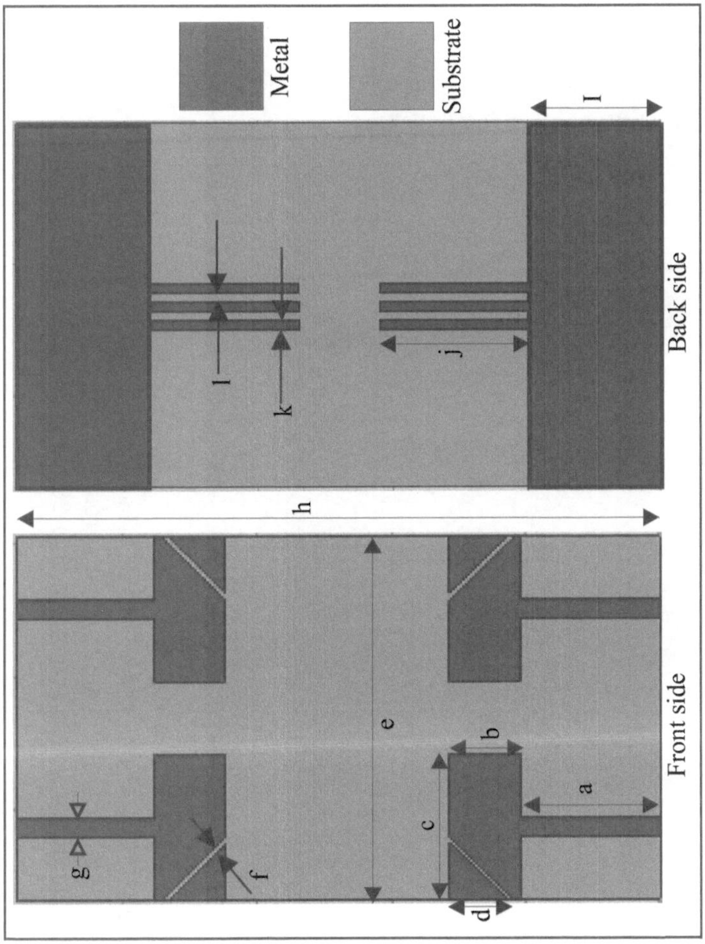

FIGURE 8.1
Schematic views of 2 × 2 MIMO antenna [208]. [From: Leeladhar Malviya, R. K. Panigrahi, and M. V. Kartikeyan, "A multi-standard, wide-band 2 × 2 compact MIMO antenna with ground modification techniques," International Journal of Microwave and Optical Technology (IJMOT), vol. 11, no. 4, pp. 259–267, July 2016. Reproduced courtesy of the International Journal of Microwave and Optical Technology.]

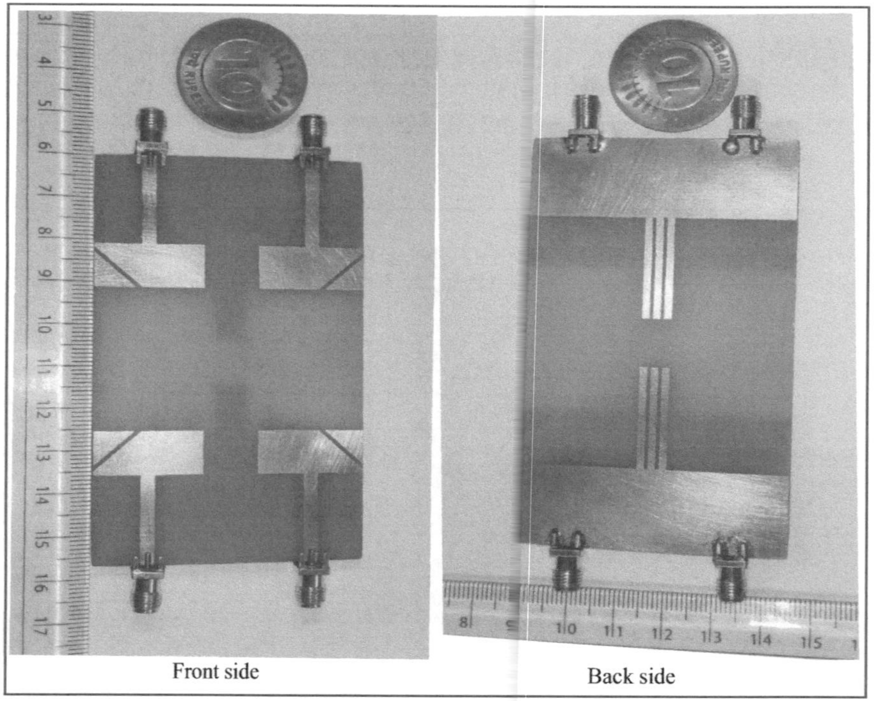

FIGURE 8.2
Fabricated views of 2×2 MIMO antenna [208]. [From: Leeladhar Malviya, R. K. Panigrahi, and M. V. Kartikeyan, "A multi-standard, wide-band 2×2 compact MIMO antenna with ground modification techniques," International Journal of Microwave and Optical Technology (IJMOT), vol. 11, no. 4, pp. 259–267, July 2016. Reproduced courtesy of the International Journal of Microwave and Optical Technology.]

represented by the parallel combination of L2-C2, S_{13} by L3-C3, and S_{14} by L4-C4. Figure 8.3 shows the equivalent circuit model of the designed wide-band MIMO. The circuit parameters are obtained as: L1 = 0.6 nH, C1 = 0.59 pf, L2 = 0.6 nH, C2 = 0.6 pf, L3 = 0.59 nH, C3 = 2.42 pf, L4 = 0.69 nH, and C4 = 1.26 pf.

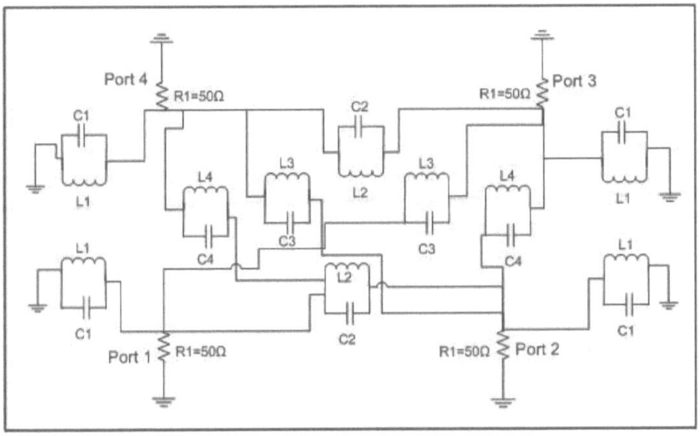

FIGURE 8.3
Equivalent circuit of wide-band MIMO [208]. [From: Leeladhar Malviya, R. K. Panigrahi, and M. V. Kartikeyan, "A multi-standard, wide-band 2×2 compact MIMO antenna with ground modification techniques," International Journal of Microwave and Optical Technology (IJMOT), vol. 11, no. 4, pp. 259–267, July 2016. Reproduced courtesy of the International Journal of Microwave and Optical Technology.]

8.3 Simulation-Measurement Results and Discussion

Wide-band MIMO antenna with optimized critical dimension parameters is designed for low return-loss and high isolation. Two-port VNA-HP8720B is calibrated to validate the S parameters, and the characterization of far-field is carried out in an anechoic chamber for the realization of CST design parameters.

Effectiveness of single element and its transformation in designed MIMO (with and without inclined slot) are shown in Fig. 8.4, along with left and right shifts. Simple patch, patch with left slot and center position, and patch with right slot and center position cover 3.5 GHz band only, and have −25 dB return-loss at resonant.

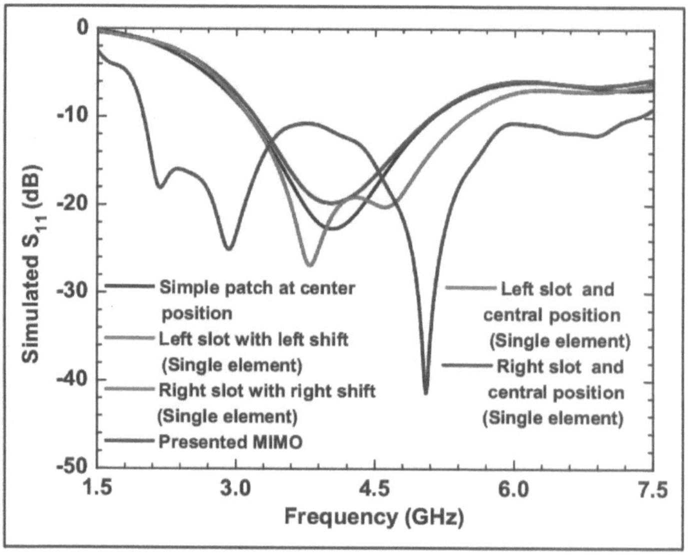

FIGURE 8.4
S_{11} of single element and wide-band MIMO (2 × 2) [208]. [From: Leeladhar Malviya, R. K. Panigrahi, and M. V. Kartikeyan, "A multi-standard, wide-band 2 × 2 compact MIMO antenna with ground modification techniques," International Journal of Microwave and Optical Technology (IJMOT), vol. 11, no. 4, pp. 259–267, July 2016. Reproduced courtesy of the International Journal of Microwave and Optical Technology.]

TABLE 8.2
Comparison of single elements and wide-band MIMO

Antenna type	Frequency band (Simulated) GHz	Bandwidth (−10 dB) GHz	Resonant frequency (f_r) GHz	Return-loss at f_r (dB)	Gain at f_r (dBi)
Simple patch (No slot) at center	3.15–5.12	1.97	4.04	−22.69	3.01
Patch with left slot and center position	3.21–5.10	1.9	4.02	−19.77	2.85
Patch with right slot and center position	3.21–5.10	1.9	4.02	−19.75	2.85
Patch with left slot and left shift	3.14–5.48	2.34	3.81	−26.88	2.25
Patch with right slot and right shift	3.14–5.48	2.34	3.81	−26.87	2.25
Presented MIMO (2 × 2)	2.01–7.31	5.30	5.05	−41.39	4.0

When patch with left slot and left shift, and patch with right slot and right shift are designed, then 3.5/5.2 GHz frequency bands are covered with −25 dB return-loss at resonant. Previous two designs are combined to implement MIMO antenna. Hence, IEEE 802.11 a/h/j/n/ac/p/y WLAN standards including ISM band, and WiMAX bands (2.5/3.5/5.5 GHz) under the frequency spectrum of 5.3 GHz, are covered. Final design verifies that the MIMO has a better response than a single antenna element. All the single elements with presented wide-band MIMO are compared in Table 8.2 with their frequency bands, bandwidth, resonant frequency, return-loss, and gain at resonant.

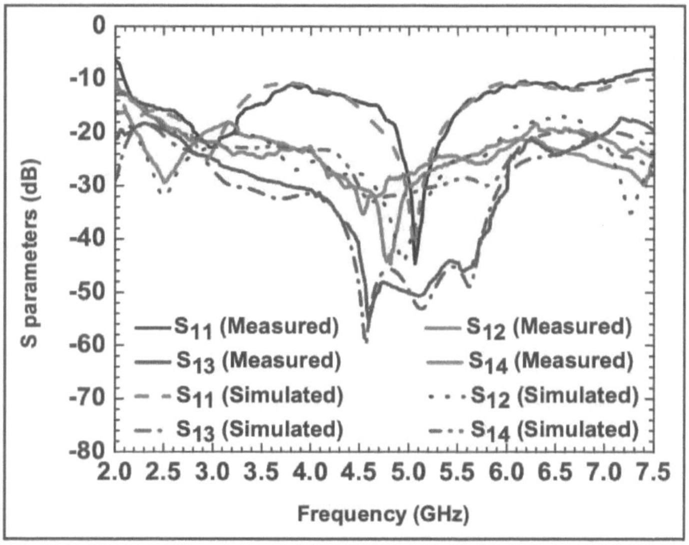

FIGURE 8.5
S parameters of 2 × 2 wide-band MIMO antenna [208]. [From: Leeladhar Malviya, R. K. Panigrahi, and M. V. Kartikeyan, "A multi-standard, wide-band 2 × 2 compact MIMO antenna with ground modification techniques," International Journal of Microwave and Optical Technology (IJMOT), vol. 11, no. 4, pp. 259–267, July 2016. Reproduced courtesy of the International Journal of Microwave and Optical Technology.]

From the simulation and measurement of *S* parameters, it is observed that $S_{11} = S_{22} = S_{33} = S_{44}$, $S_{12} = S_{21} = S_{34} = S_{43}$, $S_{13} = S_{31} = S_{24} = S_{42}$, and $S_{14} = S_{41} = S_{23} = S_{32}$. Placement of radiators with 180° phase rotated and phase reversed conditions create such situations. For easy analysis, let us consider S_{11}, S_{12}, S_{13}, and S_{14} are scattering parameters only in this chapter.

Figure 8.5 shows simulated and measured *S* parameters. Wide-band MIMO has 2:1 VSWR band of 2.01–7.31 GHz (simulated), and resonates at 5.05 GHz frequency, where −41.39 dB of return-loss is observed. At resonant

S_{12}, S_{13}, and S_{14} isolation parameters have 38.01 dB, 51.44 dB, and 31.19 dB isolations. Better than 12.6 dB isolation is achieved among adjacent and diagonal ports. Fabricated wide-band MIMO antenna has measured 2:1 VSWR band of 2.10–7.08 GHz, and resonates at 5.06 GHz frequency, where −44.56 dB return-loss is achieved. At resonant S_{12}, S_{13}, and S_{14} isolation parameters have 29.55 dB, 50.41 dB, and 27.66 dB isolations. Better than 11.0 dB isolation is achieved among adjacent and diagonal ports. Minor variation in measured and simulated responses is due to fabrication error and port/cable coupling losses.

Linking of surface current during the non-excited states (when any one port is in an excited state) reveals the amount of isolation. Figure 8.6 shows that heavy current concentrates on the microstrip feed and on the fractional ground, when port 1 is excited. Due to conduction emission (ground), some current links with port 2. However, the presence of PEG controls the linking current to port 2. Similarly, ports 3 and 4 are also linked with conduction emission of port 1. The same process is repeated to observe surface current distribution when other port is (or all are) excited. The SCD at 5.05 GHz resonant frequency has values in the range 0–25.4 A/m at different parts of radiators (due to symmetric elements the same amount is observed for different ports).

Figure 8.7 shows the effect of the full ground and the presented ground on S parameters. MIMO antenna with conventional or full ground shows 2:1 VSWR bands of 3.66–3.73 GHz and 6.80–6.96 GHz, where S_{12}, S_{13}, and S_{14} isolations are more than 23 dB in bands. Only 3.5 GHz frequency band is covered here. Wide-band MIMO with presented ground covers all the WLAN and WiMAX bands with appropriate values of isolations among ports.

Figure 8.8 shows the effect of PEG arms on S parameters. In the absence of PEG arms, 2:1 VSWR band of 2.91–6.22 GHz is covered with 9 dB isolation among ports. Although, the antenna has wide-band characteristics but unable to cover 2.4 GHz ISM band. Presented wide-band MIMO uses PEG ground arms for better coverage of all the WLAN and WiMAX bands and has better than 12.6 dB of isolation among ports.

For effectiveness of the presented wide-band MIMO antenna, partial ground length and PEG arm length are evaluated. Figure 8.9 shows the effectiveness of the length of partial ground. For the length of 16.68 mm, MIMO has 2:1 VSWR bands of 2.29–3.19 GHz and 4.47–5.63 GHz respectively. Similarly, for the length of 17.68 mm, antenna has 2:1 VSWR bands of 2.04–3.40 GHz and 3.89–5.36 GHz respectively. Considered length of the partial ground is 18.68 mm, for which the antenna shows 5.3 GHz of wide-band characteristics. For the length of 19.68 mm, the antenna also shows 2.07–7.32 GHz of wide-band. For the partial ground length of 20.68 mm, antenna resonants in 3.11–3.56 GHz and 5.68–6.02 GHz respectively. However, S_{11}, S_{12}, S_{13}, and S_{14} parameters for 18.68 mm have excellent responses.

Figure 8.10 shows S parameters for parametric evaluation on length of PEG arms. Wide-band characteristics are achieved for different lengths of

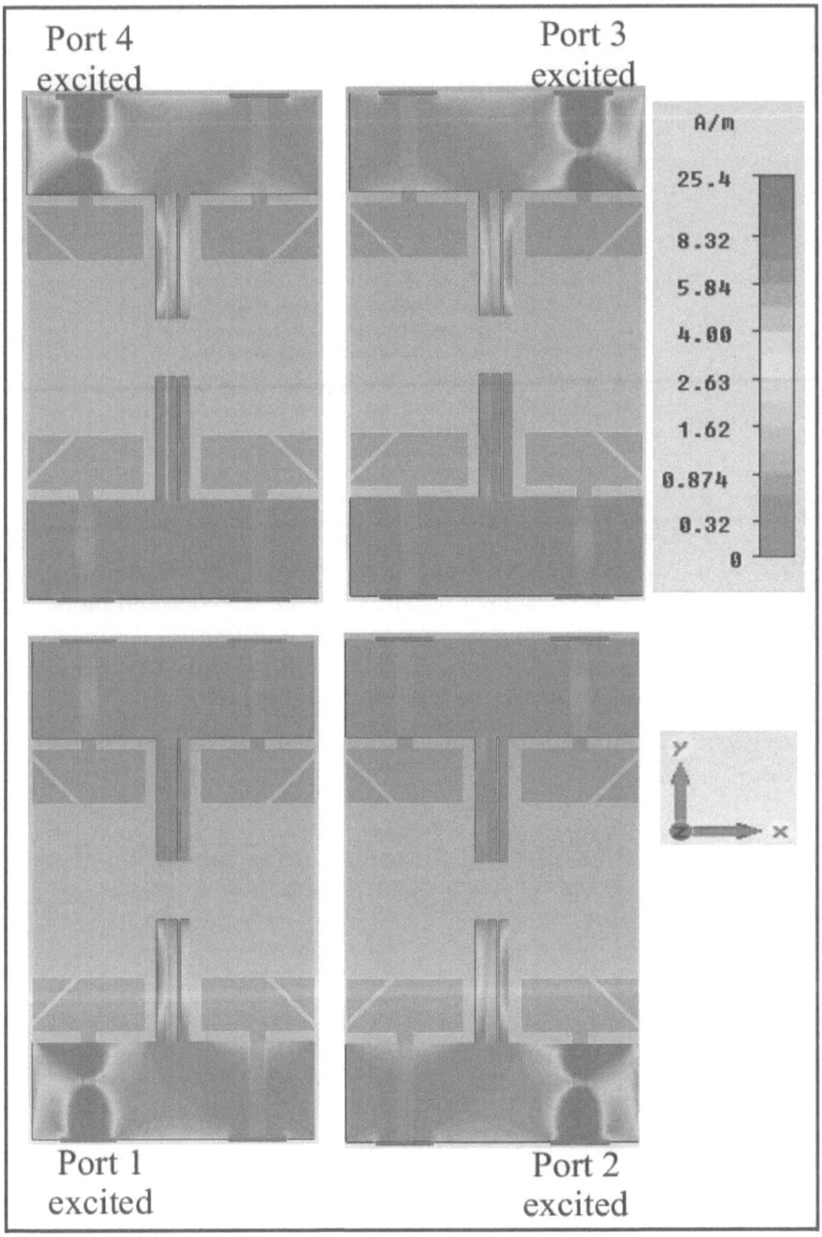

FIGURE 8.6
SCD of wide-band MIMO antenna at 5.05 GHz [208]. [From: Leeladhar Malviya, R. K. Panigrahi, and M. V. Kartikeyan, "A multi-standard, wide-band 2 × 2 compact MIMO antenna with ground modification techniques," International Journal of Microwave and Optical Technology (IJMOT), vol. 11, no. 4, pp. 259–267, July 2016. Reproduced courtesy of the International Journal of Microwave and Optical Technology.]

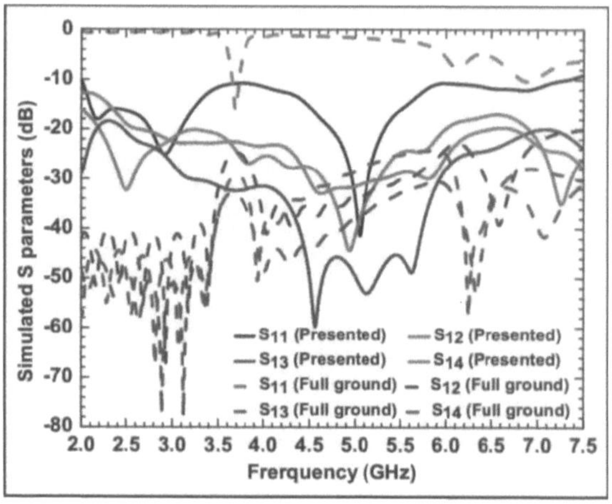

FIGURE 8.7
Comparison of wide-band MIMO with full and presented ground [208]. [From: Leeladhar Malviya, R. K. Panigrahi, and M. V. Kartikeyan, "A multi-standard, wide-band 2 × 2 compact MIMO antenna with ground modification techniques," International Journal of Microwave and Optical Technology (IJ-MOT), vol. 11, no. 4, pp. 259–267, July 2016. Reproduced courtesy of the International Journal of Microwave and Optical Technology.]

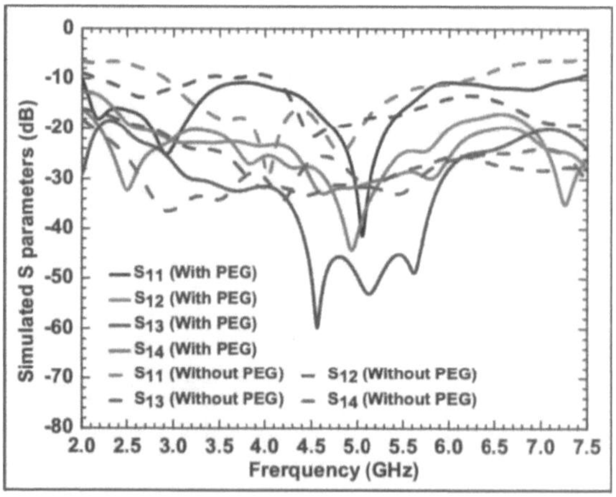

FIGURE 8.8
Comparison of wide-band MIMO with and without PEG arms [208]. [From: Leeladhar Malviya, R. K. Panigrahi, and M. V. Kartikeyan, "A multi-standard, wide-band 2 × 2 compact MIMO antenna with ground modification techniques," International Journal of Microwave and Optical Technology (IJ-MOT), vol. 11, no. 4, pp. 259–267, July 2016. Reproduced courtesy of the International Journal of Microwave and Optical Technology.]

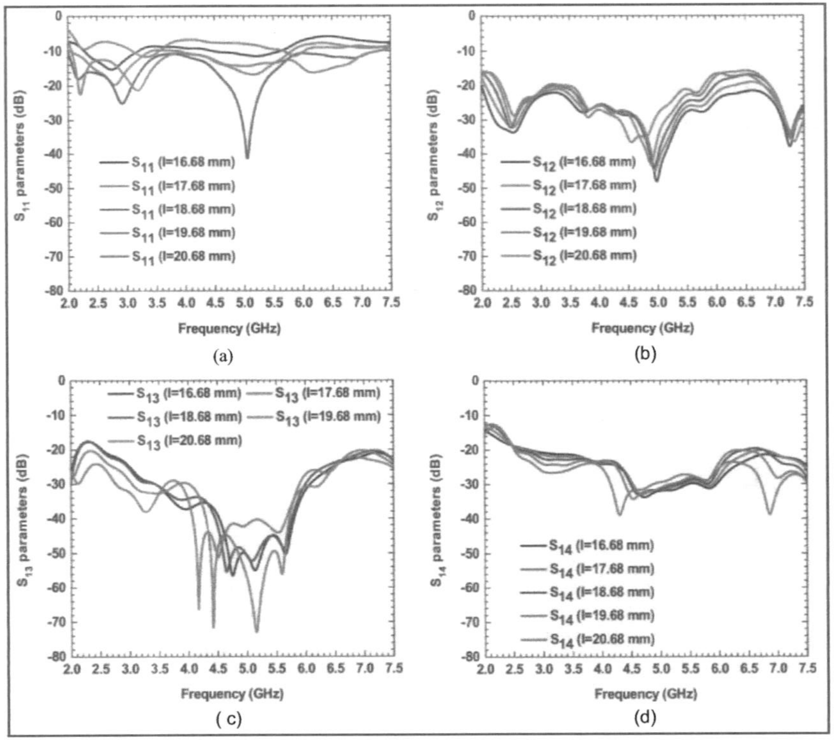

FIGURE 8.9
Effect of partial ground length on S parameters [208]. [From: Leeladhar Malviya, R. K. Panigrahi, and M. V. Kartikeyan, "A multi-standard, wide-band 2×2 compact MIMO antenna with ground modification techniques," International Journal of Microwave and Optical Technology (IJMOT), vol. 11, no. 4, pp. 259–267, July 2016. Reproduced courtesy of the International Journal of Microwave and Optical Technology.]

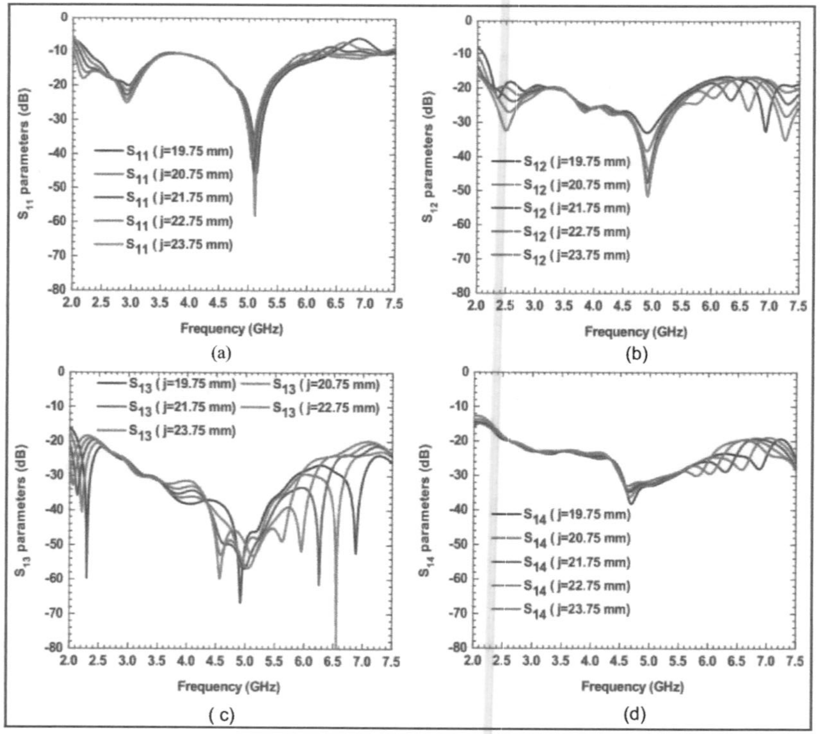

FIGURE 8.10
Effect of PEG length on S parameters [208]. [From: Leeladhar Malviya, R. K. Panigrahi, and M. V. Kartikeyan, "A multi-standard, wide-band 2×2 compact MIMO antenna with ground modification techniques," International Journal of Microwave and Optical Technology (IJMOT), vol. 11, no. 4, pp. 259–267, July 2016. Reproduced courtesy of the International Journal of Microwave and Optical Technology.]

PEG arms. Only for PEG arm of 23.75 mm, appropriate responses of S_{11}, S_{12}, S_{13}, and S_{14} are achieved. Isolation in lower-frequency band plays a vital role in multi-port antenna design. Considered length of PEG arm equals 23.75 mm justifies the selection criteria.

Similarly, Fig. 8.11 shows a comparison between simulated CST and equivalent circuit responses. The S_{11}, S_{12}, S_{13}, and S_{14} parameters are in good agreement for CST and circuit-simulated responses.

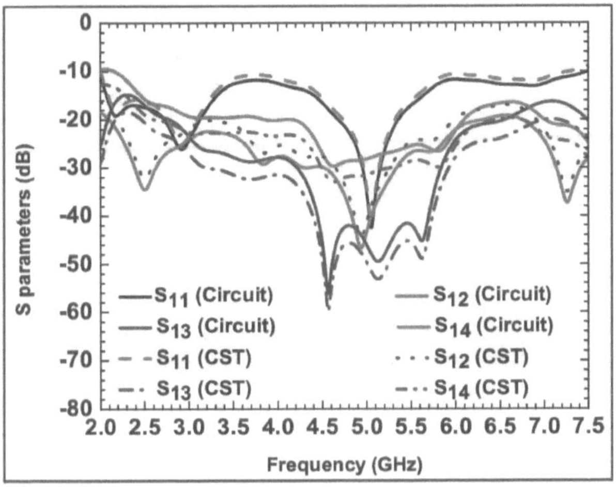

FIGURE 8.11
Circuit and CST S parameters [208]. [From: Leeladhar Malviya, R. K. Panigrahi, and M. V. Kartikeyan, "A multi-standard, wide-band 2 × 2 compact MIMO antenna with ground modification techniques," International Journal of Microwave and Optical Technology (IJMOT), vol. 11, no. 4, pp. 259–267, July 2016. Reproduced courtesy of the International Journal of Microwave and Optical Technology.]

For the wide-band MIMO, diversity performances are described using MIMO-gain, efficiency, ECC, TARC, and MEG. For single element (for all the single element stages) simulated gain is in the range of 4.0 dBi. Figure 8.12 shows that wide-band MIMO antenna has measured gain in the range 1.86–6.26 dBi, and simulated gain in the range of 1.87–6.40 dBi in the given frequency band. Also, 3.7 dBi measured gain at 5.06 GHz, and 4.0 dBi simulated gain at 5.05 GHz are achieved. Due to widening in bandwidth, the gain has limited value at the lower-frequency side. The radiation and total efficiencies for the considered single radiating element are in the range of 90%. Similarly, wide-band MIMO radiation efficiency in the frequency range is 62.95%–94.06% (simulated), and 87.74% at resonant. Total efficiency for wide-band MIMO ranges between 56.0 and 91.57% and is 87.66% at resonant. Hence, the wide-band MIMO has much better gain and efficiency responses in comparison with considered single element.

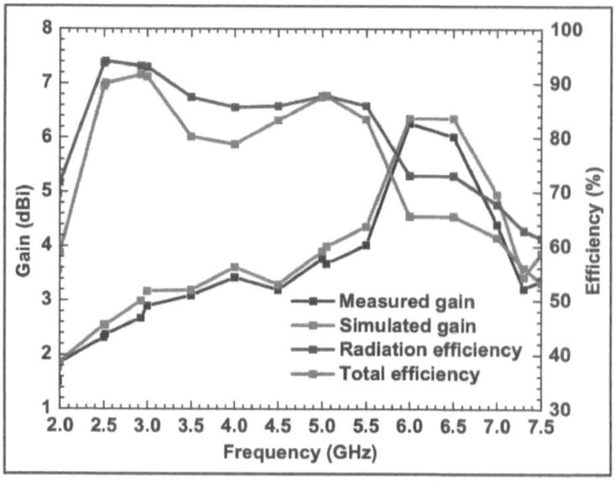

FIGURE 8.12
Gain and efficiency of wide-band MIMO antenna [208]. [From: Leeladhar Malviya, R. K. Panigrahi, and M. V. Kartikeyan, "A multi-standard, wide-band 2 × 2 compact MIMO antenna with ground modification techniques," International Journal of Microwave and Optical Technology (IJMOT), vol. 11, no. 4, pp. 259–267, July 2016. Reproduced courtesy of the International Journal of Microwave and Optical Technology.]

Wide-band MIMO diversity performance in terms of ECC may be obtained by substituting the terms $S_{11} = S_{22} = S_{33} = S_{44}$, $S_{12} = S_{21} = S_{34} = S_{43}$, $S_{14} = S_{41} = S_{23} = S_{32}$, and $S_{13} = S_{31} = S_{24} = S_{42}$ in (4.7). Here the values of $i =$ 1 to 2, $j = 1$ to 2, for any two elements, and $N = 2$, as 2 antenna elements are selected at a time. ECC between different ports is represented by ρ_{e12}, ρ_{e13}, and ρ_{e14}. Figure 8.13 shows that simulated ECC in the considered frequency range is 0–0.02, and measured ECC is in the range 0–0.2 for fabricated MIMO. At 5.06 GHz resonant frequency, the simulated ECC values are $\rho_{e12} = 7.46 \times 10^{-8}$, $\rho_{e13} = 1.74 \times 10^{-9}$, and $\rho_{e14} = 4.4 \times 10^{-8}$. Similarly, the measured ECC values at 5.06 GHz frequency are $\rho_{e12} = 7.1 \times 10^{-2}$, $\rho_{e13} = 8.1 \times 10^{-2}$, and $\rho_{e14} = 8.2 \times 10^{-2}$. Large fluctuations in the measured ECCs are due to the fabrication errors and port/cable coupling losses. Even, measured ECCs are in the range of 0.2 for the considered frequency spectrum, and satisfy the criteria of wireless MIMO designs.

Diversity performance of wide-band MIMO in terms of TARC is obtained using (4.10). Figure 8.14 shows the best excitation angle combination of 90°, 180° between adjacent and diagonal ports, where 2:1 VSWR bandwidth is more than 5.0 GHz. Other phase combinations show distortions in TARC responses. TARC responses resemble the shape of obtained isolation parameters of wide-band MIMO.

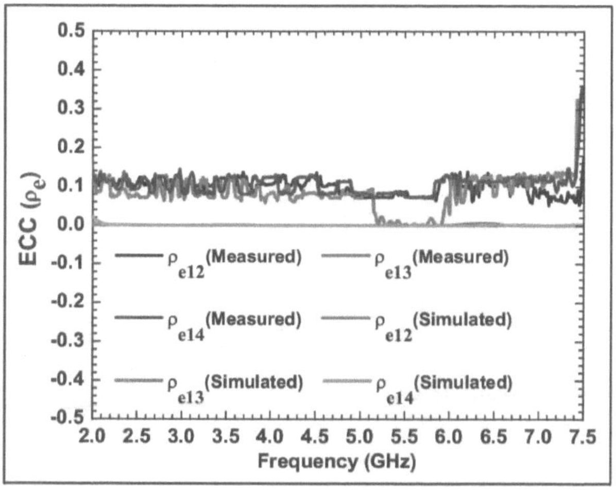

FIGURE 8.13
ECC of wide-band MIMO antenna [208]. [From: Leeladhar Malviya, R. K. Panigrahi, and M. V. Kartikeyan, "A multi-standard, wide-band 2×2 compact MIMO antenna with ground modification techniques," International Journal of Microwave and Optical Technology (IJMOT), vol. 11, no. 4, pp. 259–267, July 2016. Reproduced courtesy of the International Journal of Microwave and Optical Technology.]

Wide-band MIMO diversity behavior in terms of MEG can be obtained by (4.15). Let the Gaussian medium signals have mean (μ) = 0 and variance (σ) = 20, for horizontal and vertical components. Figure 8.15 shows the responses of MEGs with Isotropic and Gaussian mediums for different XPR values using CST simulations. MEG for an Isotropic medium with $XPR =$ 0 dB, lies in the range of −4.7 to −6.7 dB, and for $XPR =$ 6 dB, MEG lies in the range of −6.0 to −8.0 dB. Similarly, for Gaussian medium with XPR = 0 dB, MEG lies in the range of −5.6 to −7.2 dB, and for $XPR =$ 6 dB, MEG lies in the range of −8.4 to −10.0 dB. Hence, wide-band MIMO antenna shows strong candidature for both the indoor and outdoor activities for both the Isotropic and Gaussian mediums.

Wide-band MIMO antenna characteristics in far-field are measured in the anechoic chamber at 5.06 GHz resonant frequency, in presence of standard transmitting horn antenna. Wide-band MIMO radiators are used in receiving mode to get E-field and H-field radiation characteristics. During the measurement, all unused ports are terminated with 50.0 Ω to avoid any noise pick-up.

Figure 8.16 shows the simulated E-field and H-field radiation patterns of single elements (patch with left slot and left shift from center position, and patch with right slot and right shift from center position). These two single elements are transformed into the designed wide-band MIMO. Left and right shifts with the partial ground in single element show the variation in

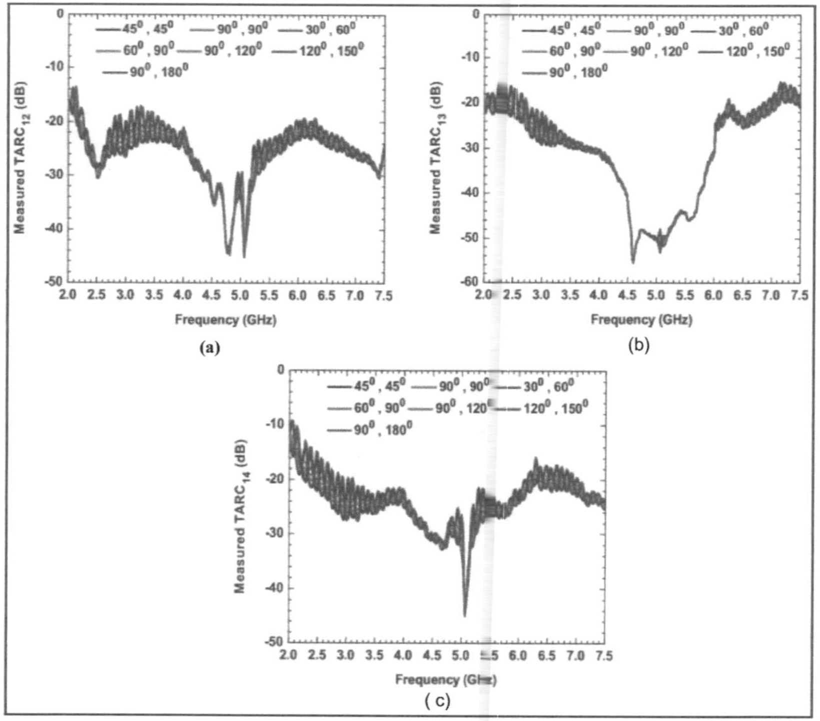

FIGURE 8.14
TARC of wide-band MIMO antenna [208]. [From: Leeladhar Malviya, R. K. Panigrahi, and M. V. Kartikeyan, "A multi-standard, wide-band 2×2 compact MIMO antenna with ground modification techniques," International Journal of Microwave and Optical Technology (IJMOT), vol. 11, no. 4, pp. 259–267, July 2016. Reproduced courtesy of the International Journal of Microwave and Optical Technology.]

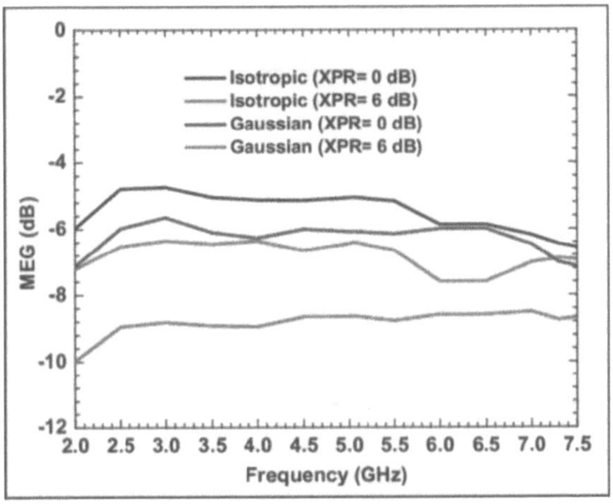

FIGURE 8.15
MEG of wide-band MIMO antenna [208]. [From: Leeladhar Malviya, R. K. Panigrahi, and M. V. Kartikeyan, "A multi-standard, wide-band 2×2 compact MIMO antenna with ground modification techniques," International Journal of Microwave and Optical Technology (IJMOT), vol. 11, no. 4, pp. 259–267, July 2016. Reproduced courtesy of the International Journal of Microwave and Optical Technology.]

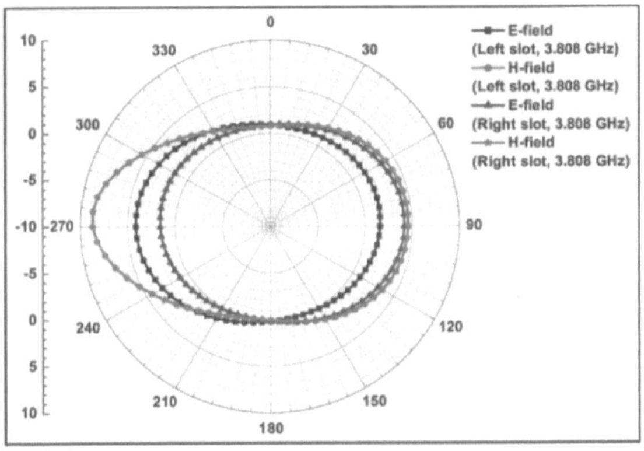

FIGURE 8.16
Far-field patterns of single element [208]. [From: Leeladhar Malviya, R. K. Panigrahi, and M. V. Kartikeyan, "A multi-standard, wide-band 2×2 compact MIMO antenna with ground modification techniques," International Journal of Microwave and Optical Technology (IJMOT), vol. 11, no. 4, pp. 259–267, July 2016. Reproduced courtesy of the International Journal of Microwave and Optical Technology.]

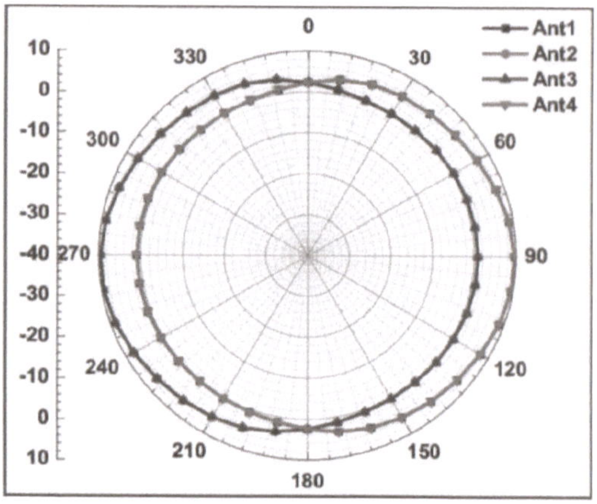

FIGURE 8.17
Wide-band MIMO E-field patterns at 5.06 GHz (simulated) [208]. [From: Leeladhar Malviya, R. K. Panigrahi, and M. V. Kartikeyan, "A multi-standard, wide-band 2 × 2 compact MIMO antenna with ground modification techniques," International Journal of Microwave and Optical Technology (IJ-MOT), vol. 11, no. 4, pp. 259–267, July 2016. Reproduced courtesy of the International Journal of Microwave and Optical Technology.]

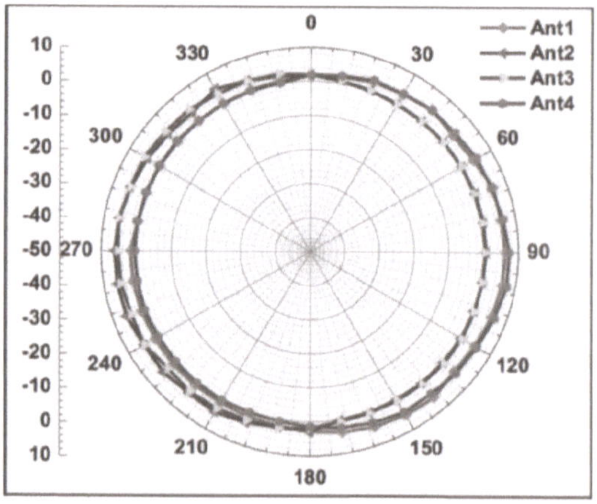

FIGURE 8.18
Wide-band MIMO E-field patterns at 5.06 GHz (measured) [208]. [From: Leeladhar Malviya, R. K. Panigrahi, and M. V. Kartikeyan, "A multi-standard, wide-band 2 × 2 compact MIMO antenna with ground modification techniques," International Journal of Microwave and Optical Technology (IJ-MOT), vol. 11, no. 4, pp. 259–267, July 2016. Reproduced courtesy of the International Journal of Microwave and Optical Technology.]

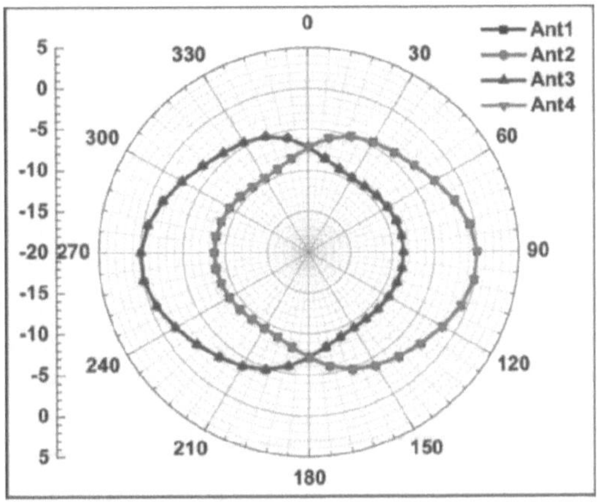

FIGURE 8.19
Wide-band MIMO H-field patterns at 5.06 GHz (simulated) [208]. [From: Leeladhar Malviya, R. K. Panigrahi, and M. V. Kartikeyan, "A multi-standard, wide-band 2 × 2 compact MIMO antenna with ground modification techniques," International Journal of Microwave and Optical Technology (IJMOT), vol. 11, no. 4, pp. 259–267, July 2016. Reproduced courtesy of the International Journal of Microwave and Optical Technology.]

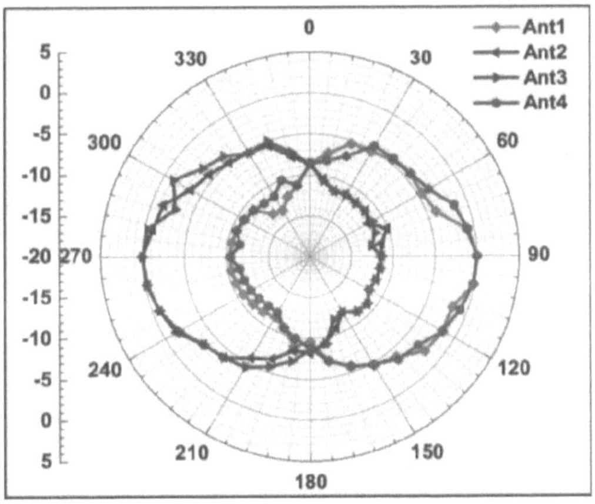

FIGURE 8.20
Wide-band MIMO H-field patterns at 5.06 GHz (measured) [208]. [From: Leeladhar Malviya, R. K. Panigrahi, and M. V. Kartikeyan, "A multi-standard, wide-band 2 × 2 compact MIMO antenna with ground modification techniques," International Journal of Microwave and Optical Technology (IJMOT), vol. 11, no. 4, pp. 259–267, July 2016. Reproduced courtesy of the International Journal of Microwave and Optical Technology.]

both E-field and H-field at 3.81 GHz resonant frequency. Diversity characterization may be seen with these two shifts of a single element. Designed wide-band MIMO shows exact complementary radiation patterns. These are the combined effect of placement of radiators. Also, it is verified that wide-band MIMO has much better far-field characteristics than a single radiating element.

Figures 8.17 and 8.18 show simulated and measured E-field patterns. E-field main lobe directions of radiating elements 1 and 4 is 325°, and for 2 and 3 elements is 35°. Similarly, Figs. 8.19 and 8.20 show simulated and measured H-field patterns. H-field main lobe directions for radiating elements 1 and 2 are 335°, and for 3 and 4 elements is 25°. Variations in these far-field patterns are due to the fabrication errors and port coupling losses.

8.4 Concluding Remarks

Wide-band MIMO antenna with diversity effects has been presented. Joint operation of partial ground and PEG arms contributed in compactness, wide-banding, and enhanced isolation. Wide-band MIMO antenna design covered 2.1–7.09 GHz measured band under 2:1 VSWR. Better than 11.0 dB of isolation, coverage of IEEE 802.11 a/h/j/n/ac/p/y WLAN standards, and 2.5/3.5/5.5 GHz WiMAX bands are the advantages of designed wide-band MIMO. Small value of ECC, wide TARC bandwidth, and applicability of wide-band MIMO in indoor and outdoor Gaussian environment makes it suitable for wireless applications [208].

9

Design and Analysis of CP-MIMO Antenna for WLAN Application

Linearly polarized antennas in line-of-sight communication are able to receive un-equal signal powers. Receiving antenna with the lowest power places the higher limit on diversity gain and in turn on signal-to-noise ratio (SNR). Also, due to polarization mismatch, highly isolated antennas receive inadequate signal gains in indoor and outdoor environments. In recent years, lots of research and industrial evolution have witnessed the overwhelming relevance of circularly polarized (CP) antennas in wireless activities. CP antennas may equally divide signal powers among receiving antennas, by solving the problem of polarization mismatch. CP-MIMO antennas enhance the data rate, capacity, and diversity gain. This chapter focuses on the design of 2×2 CP-MIMO antenna to resonate at 5.8 GHz IEEE 802.11 WLAN band for non-line-of-sight (NLOS) communication.

9.1 Introduction and Related Work

Linearly polarized microstrip antennas can be transformed in CP antennas, with perturbation in feeding approach and some specific conditions. Such perturbations excite the E and H fields to generate left hand (LH) and right hand (RH) polarized waves that are called as left hand circular polarization (LHCP) and right hand circular polarization (RHCP). The CP antenna has focusing, anti-jamming, and anti-interference capabilities. Circular polarization (CP) solves multipath propagation problems in all environmental conditions, irrespective of the type of fading and diversity techniques. It adds the flexibility in phase variations and makes the antennas orientation independent. Therefore, radiations of CP antenna penetrate in all directions (vertical as well as in horizontal) [209–213].

Various approaches of CP antenna designs are described here. First approach of CP generation concentrates on diagonal axis and coaxial feed with conventional microstrip patches to create two orthogonal modes for CP generation. In the second approach, a conventional patch with truncated corners or slots with coaxial feed generates two orthogonal modes for CP wave generation [214]. In the third approach, 45° tilted slot at the center of the patch and with microstrip feed generates CP waves i.e. TM_{01} and TM_{10} modes. CP antenna design with microstrip feed avoids the cutting of substrate and conducting layers of the radiating structure, and in turn, controls the secondary sources of radiation emission [215]. In the fourth approach, the square slot with unequal L-shaped patch/ground arms [216], and asymmetric slits are used to generate CP mode [217]. These approaches may be combined to create CP polarization.

CP mode can be generated in focused antenna arrays using L-shaped slot [218], and also by rotating slots to transfer maximum power [219]. Segmented circles with orthogonal phase coupling and divider network can produce CP wave for MIMO antennas [220, 221]. Multi-layer substrates with orthogonal feed structure can produce equal amplitudes and broadband CP antennas with high gain [222, 223]. Proximity and gap coupled multi-feed approaches are required for high gain antennas for beam steering and high front-to-back lobe ratio (FBR) [224,225]. The effect of surface waves in antenna arrays limit the gain. This effect can be limited by using shorting pins/posts at diagonal position [226], ground separator, and by orientations of radiators [227].

In this chapter, CP-MIMO antenna with LHCP and RHCP resonates at 5.8 GHz IEEE 802.11 WLAN standard, has more than 33.0 dB isolation, and axial ratio (AR) $\leq$ 3.0 dB.

9.2 CP-MIMO Antenna Design and Implementation

Two-port LHCP and RHCP MIMO antenna with 50.0 Ω ports are designed to uniformly distribute signal powers between radiators. FR-4 dielectric substrate (thickness of 1.524 mm, permittivity of 4.4, and loss tangent of 0.025) of size 27.69 × 97.0 mm^2 is used for the fabrication. Microstrip feed is preferred to avoid the emission of secondary fields.

CP in presented 2 × 2 MIMO antenna is generated by using 90° apart rectangular slots at the center of a rectangular patch. Two orthogonal modes are generated with 90° phase and equal amplitudes and are excited by the slots. Total power is divided between radiators using 1 × 2 feed mechanism (a combination of 50.0 Ω, 70.7 Ω, and 100.0 Ω microstrip lines). Full ground with 0.5 mm ground split, and 1.5 mm slot between the radiating elements of 1 × 2 power divider arm controls the conduction emission and inter-element coupling.

The frequency of operation = 5.8 GHz (5.725–5.825 GHz), VSWR = 2:1, isolation >30 dB, gain (dBi) >5, efficiency (%) >90, ECC <0.1, circular polarization, axial ratio ≤1, and WLAN application are the set goals of the MIMO antenna design here. Figures 9.1 and 9.2 show the schematic and fabricated views of CP-MIMO antenna. The optimized parameters are given in Table 9.1.

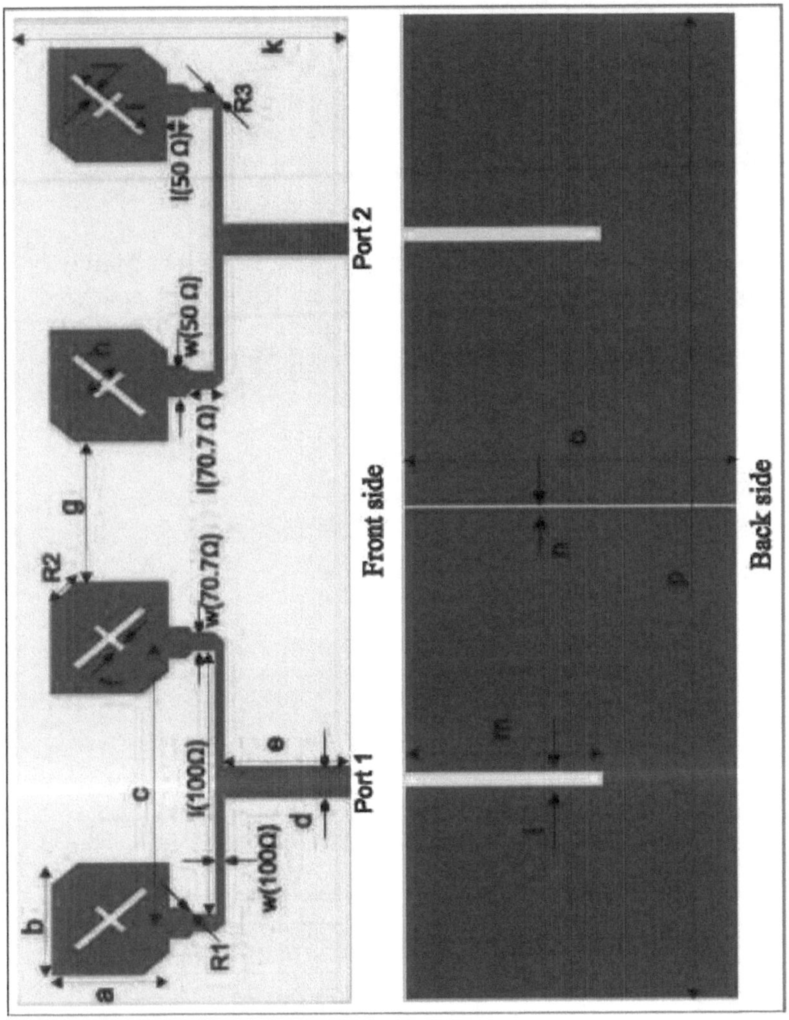

FIGURE 9.1
Schematic views of 2 × 2 CP-MIMO antenna [228]. [From: Leeladhar Malviya, Rajib K. Panigrahi, and Machavaram V. Kartikeyan, "Circularly polarized 2 × 2 MIMO antenna for WLAN applications," Progress in Electromagnetics Research C, vol. 66, pp. 97–107, July 2016. Reproduced courtesy of the Electromagnetics Academy.]

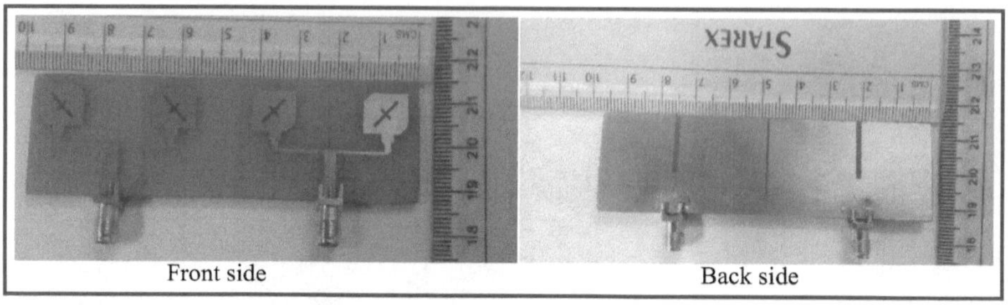

FIGURE 9.2
Fabricated views of 2 × 2 CP-MIMO antenna [228]. [From: Leeladhar Malviya, Rajib K. Panigrahi, and Machavaram V. Kartikeyan, "Circularly polarized 2 × 2 MIMO antenna for WLAN applications," Progress in Electromagnetics Research C, vol. 66, pp. 97–107, July 2016. Reproduced courtesy of the Electromagnetics Academy.]

TABLE 9.1
CST optimized parameters of CP-MIMO antenna (Unit: mm) [228]. [From: Leeladhar Malviya, Rajib K. Panigrahi, and Machavaram V. Kartikeyan, "Circularly polarized 2 × 2 MIMO antenna for WLAN applications," Progress in Electromagnetics Research C, vol. 66, pp. 97–107, July 2016. Reproduced courtesy of the Electromagnetics Academy.]

Parameter	a	b	c	e	f
Value	9.64	11.0	24.98	10.5	0.692
Parameter	g	h	l(50 Ω)	d = w(50 Ω)	l(70.7 Ω)
Value	13.75	3.26	2.0	2.96	2.0
Parameter	i	j	k = o	l	m
Value	8.12	0.74	27.69	1.5	16.5
Parameter	n	p	w(70.7 Ω)	l(100 Ω)	w(100 Ω)
Value	0.5	97.0	1.58	26.12	0.694

Figure 9.3 shows the CP-MIMO equivalent circuit diagram, developed using ADS software for tuning behavior of patch, slot, and mutual coupling. Each port is modeled using 50.0 Ω load. Feed structure (50.0 Ω, 70.7 Ω, and 100.0 Ω microstrip transmission lines) is represented by the parallel combination of L1 and C1. Microstrip patch is represented by the parallel combination of L2 and C2, effect of two asymmetric slots by the parallel combination of L3 and C3, and mutual coupling by the parallel combination of L4 and C4. The circuit elements L1 = 0.9 nH, L2 = 0.66 nH, L3 = 1.2 nH, L4 = 1.99 nH, C1 = 0.6 pF, C2 = 0.76 pF, C3 = 0.86 pF, and C4 = 0.9 pF are achieved using ADS software.

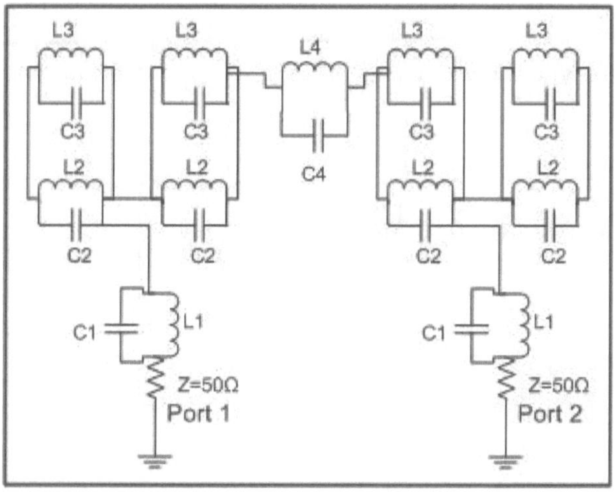

FIGURE 9.3
Equivalent circuit of CP-MIMO [228]. [From: Leeladhar Malviya, Rajib K. Panigrahi, and Machavaram V. Kartikeyan, "Circularly polarized 2×2 MIMO antenna for WLAN applications," Progress in Electromagnetics Research C, vol. 66, pp. 97–107, July 2016. Reproduced courtesy of the Electromagnetics Academy.]

9.3 Simulation-Measurement Results and Discussion

Testing of *S* parameters is performed with calibrated two-port VNA-HP8720B, and far-field characterization is carried out in an anechoic chamber for the validity of CP-MIMO design parameters. Three cases of patch structure are considered in Fig. 9.4 and Fig. 9.5, for return-loss, isolation and AR results.

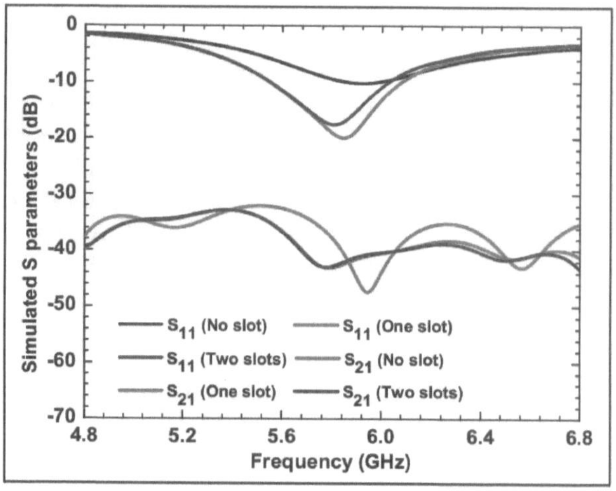

FIGURE 9.4
S parameters for different MIMO cases [228]. [From: Leeladhar Malviya, Rajib K. Panigrahi, and Machavaram V. Kartikeyan, "Circularly polarized 2×2 MIMO antenna for WLAN applications," Progress in Electromagnetics Research C, vol. 66, pp. 97–107, July 2016. Reproduced courtesy of the Electromagnetics Academy.]

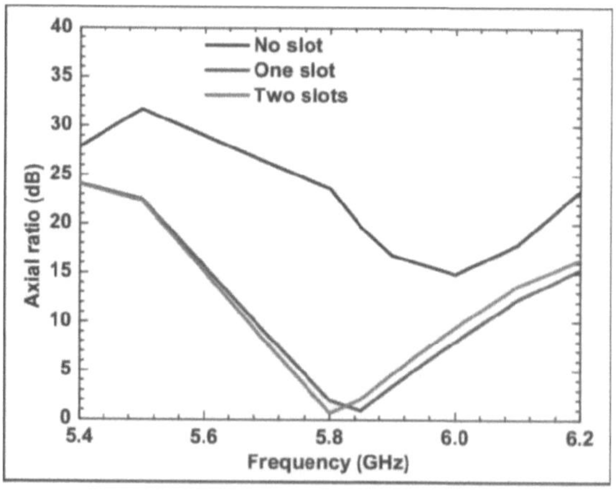

FIGURE 9.5
Axial ratio for different MIMO cases (Simulated) [228]. [From: Leeladhar Malviya, Rajib K. Panigrahi, and Machavaram V. Kartikeyan, "Circularly polarized 2×2 MIMO antenna for WLAN applications," Progress in Electromagnetics Research C, vol. 66, pp. 97–107, July 2016. Reproduced courtesy of the Electromagnetics Academy.]

In the first case, MIMO antenna with 2:1 VSWR band of 5.82–5.99 GHz frequency range is achieved with no slot in the patch. In this case, MIMO polarization is LP as AR is greater than 3.0 dB, bandwidth is 174.0 MHz, and in-band isolation is better than 35.0 dB. MIMO antenna resonates here at 5.91 GHz frequency. In the first case, return-loss is very close to −10 dB. Hence, trimming of patch and feed structure is treated for afterward two cases to reduce return-loss. In the second case, CP mode is generated using one inclined slot of 45° in the patch. MIMO antenna with 2:1 VSWR band of 5.55–6.08 GHz frequency range is achieved. More than 33.0 dB of isolation, AR of 0.93 dB at 5.85 GHz resonant frequency and spectrum shift of 672.0 MHz is achieved here in contrast with the first case. MIMO antenna bandwidth in this case is 528.0 MHz. At 5.8 GHz frequency, AR is 1.95 dB.

In the third case, the patch with two 90° apart asymmetric slots is considered for better AR characteristics. MIMO antenna with 2:1 VSWR band of 5.49–6.02 GHz frequency range is achieved here. More than 33.0 dB of isolation, AR of 0.68 dB at 5.8 GHz resonant frequency and spectrum shift of 6.0 MHz is achieved here in contrast with the second case. MIMO antenna bandwidth, in this case, is 534.0 MHz. The second slot is responsible for the extension of the band and better AR results.

The CST simulated parameters and VNA measurements are in good agreement. From the simulation and measurement of S parameters, $S_{11} = S_{22}$ and $S_{12} = S_{21}$. This is possible due to the placement of radiators in mirror position. Therefore, S_{11} and S_{21} scattering parameters are sufficient for easy analysis of results, throughout the chapter.

Figure 9.6 shows measured and simulated S parameters of the designed MIMO. All the results are corresponding to the third case discussed in the previous paragraph. The simulated 3.0 dB AR band covers 5.77–5.86 GHz frequency range, where AR bandwidth is 92.0 MHz. The measured 2:1 VSWR band (−10 dB impedance band) is 5.56–6.01 GHz, where isolation is more than 34.0 dB. At 5.8 GHz resonant frequency, isolation and AR values are 40.95 dB and 0.98 dB. The measured 3.0 dB AR band covers 5.74–5.83 GHz and provides 86.0 MHz AR bandwidth. Figure 9.7 shows all the AR results of designed CP-MIMO.

Figure 9.8 shows the AR measurement of fabricated CP-MIMO antenna set-up in the anechoic chamber in the presence of standard dual ridge horn (DRH)-10 antenna. Fabricated CP-MIMO antenna is connected to the microwave generator in transmitting mode for AR measurement, and DRH-10 is used as a receiving antenna and is connected to the power meter for tracing and recording the received power for different angular positions. Transmitter and receiver are separated by 1.5-meter distance. The measured power is then plotted for each angle on the polar plot to get the major and minor axis for AR measurement (AR is the ratio of major to minor axis powers).

Placement of CP-MIMO antenna elements leads to a high value of isolation ($S_{21} = S_{12}$) between ports. More than 33.0 dB of isolation is the result of 1×2 feed structure, the slot between feed arms, and ground split. Feeder arms are

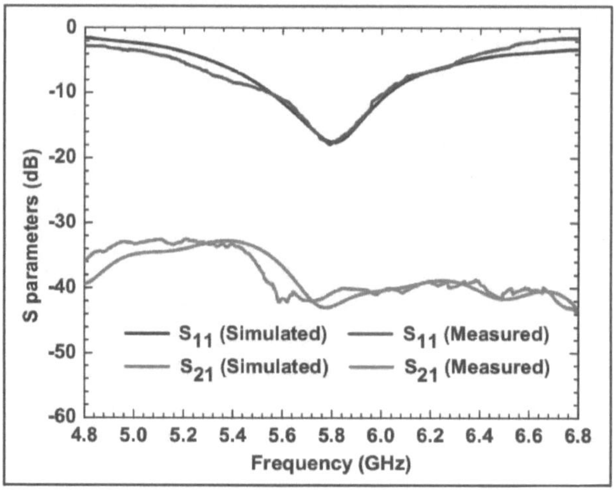

FIGURE 9.6
S parameters of designed MIMO [228]. [From: Leeladhar Malviya, Rajib K. Panigrahi, and Machavaram V. Kartikeyan, "Circularly polarized 2 × 2 MIMO antenna for WLAN applications," Progress in Electromagnetics Research C, vol. 66, pp. 97–107, July 2016. Reproduced courtesy of the Electromagnetics Academy.]

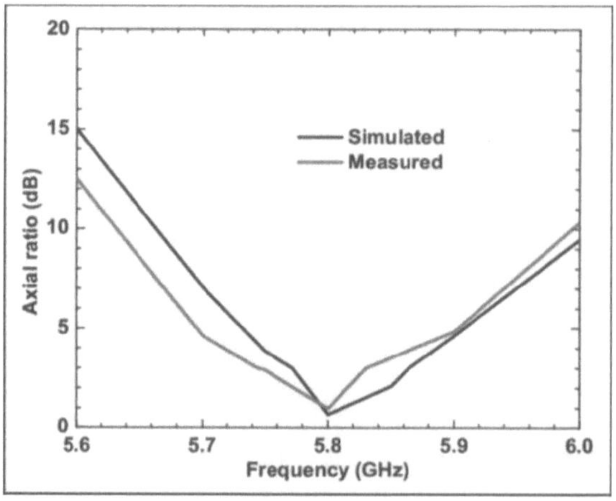

FIGURE 9.7
Axial ratio of designed MIMO [228]. [From: Leeladhar Malviya, Rajib K. Panigrahi, and Machavaram V. Kartikeyan, "Circularly polarized 2 × 2 MIMO antenna for WLAN applications," Progress in Electromagnetics Research C, vol. 66, pp. 97–107, July 2016. Reproduced courtesy of the Electromagnetics Academy.]

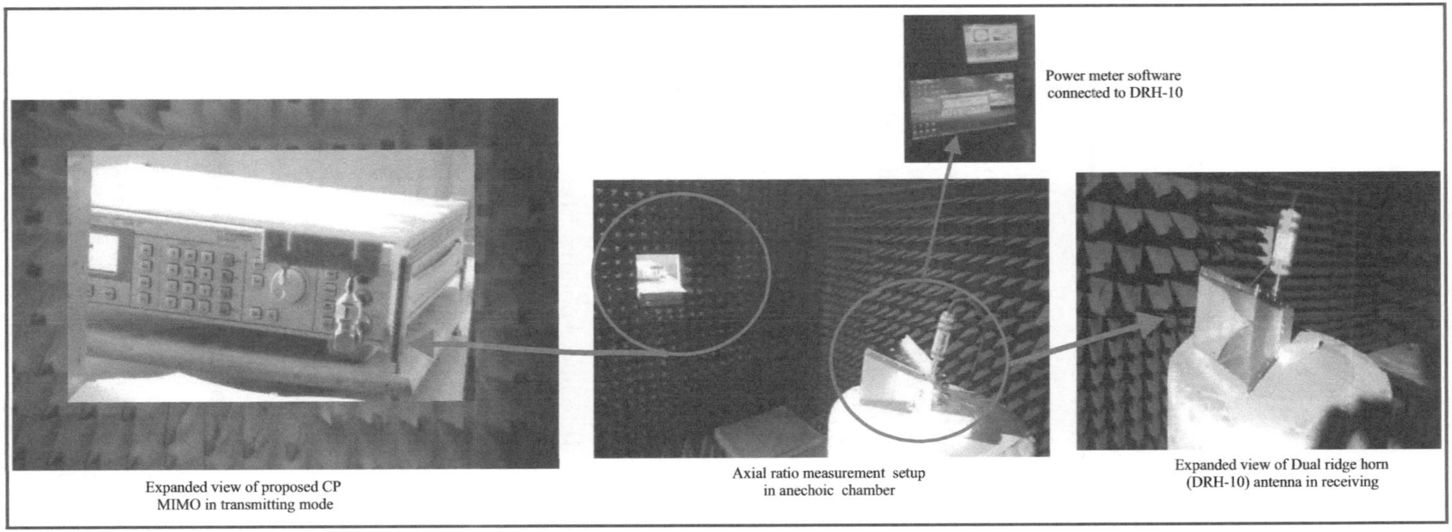

FIGURE 9.8
Axial ratio measurement set-up [228]. [From: Leeladhar Malviya, Rajib K. Panigrahi, and Machavaram V. Kartikeyan, "Circularly polarized 2 × 2 MIMO antenna for WLAN applications," Progress in Electromagnetics Research C, vol. 66, pp. 97–107, July 2016. Reproduced courtesy of the Electromagnetics Academy.]

separated by $\lambda/2$ wavelength for proper operation and desired CP-MIMO performance characteristics. A slot is etched in the ground between feeder arms to control inter-element coupling, and the ground split controls direct coupling. Due to such arrangement, low surface current concentrates at non-excited elements. Figure 9.9 shows the major current on the feed line and radiating patch, and minor current links to the non-excited element when port 1 or port 2 is excited. Under the excitation state of both the ports, major current concentrates on each of the excited ports and corresponding radiating elements.

Figures 9.10 and 9.11 show S parameters and AR results with and without the slot above the feed line. Slot etching above the feed line improves isolation by 3.44 dB at resonant. MIMO without the slot above feed line shows 2:1 VSWR band of 5.50–6.02 GHz, and resonants at 5.82 GHz. At resonant, isolation is 38.25 dB. Better return-loss and more than 33.0 dB of isolation characteristics are observed in the considered band here. Without the slot above feed line has a high impact on AR values. This slot may change the polarization also. Designed CP-MIMO uses the slot above feed line to get better than 33.0 dB in-band isolation, and AR value of 0.68 dB (as explained in the third case and measured S parameters). Hence, the slot above the feed line has a positive impact on CP (and on AR).

Effect of phase difference (θ_1^0) between the major and minor slots may be studied to find its impact on AR and VSWR. Value of θ_1^0 is varied from 0° to 90°. Major and minor slots are merged for 0° phase difference. In this case, the value of AR is 1.95 dB, and VSWR is 1.11. For 90° phase difference between slots, AR is 0.68 dB, and VSWR is 1.12. For 45° phase difference between slots, MIMO antenna has an AR value of 40.0 dB, and VSWR is 2.57. A major difference in AR values is observed at 5.8 GHz resonant frequency. Hence, for 0° and 90° of phase difference between the slots, CP characteristics are observed. Other values of θ_1^0 leads to linear polarization. The set of asymmetric slots with θ_1^0 equals 90° shows better AR characteristics than any other cases. Figure 9.12 shows the comparison of these values.

Similarly, the CST and equivalent circuit simulated responses in terms of S parameters are shown in Fig. 9.13. The S parameter responses using equivalent circuit and CST-MWS have very minor differences.

The diversity performance of 2×2 CP-MIMO antenna is characterized in terms of gain, ECC, TARC, and MEG here. At 5.8 GHz resonant frequency, the simulated gain is 5.34 dBi and measured gain is 5.23 dBi. Radiation efficiency in the presented band is better than 56.0%, and 57.92% at resonant. Total efficiency is more than 49.0% in the band. Figure 9.14 shows the comparison of gain and efficiencies. Similarly, left hand polarization of antenna 1 and right hand polarization of antenna 2 are shown in Figs. 9.15 and 9.16 for the effectiveness of CP-MIMO antenna.

ECC may be obtained using (4.7) for the designed CP-MIMO antenna. Figure 9.17 shows that the simulated ECC lies in the range 0–0.005, which is very close to zero axis in the whole band. The measured ECC lies in the range of 0–0.15. Simulated and measured ECCs at 5.8 GHz resonant frequency are 2.39×10^{-7} and 2.6×10^{-2}. In measured ECC, certain fluctuation are observable, though, in the whole band, it is also very low. The difference in

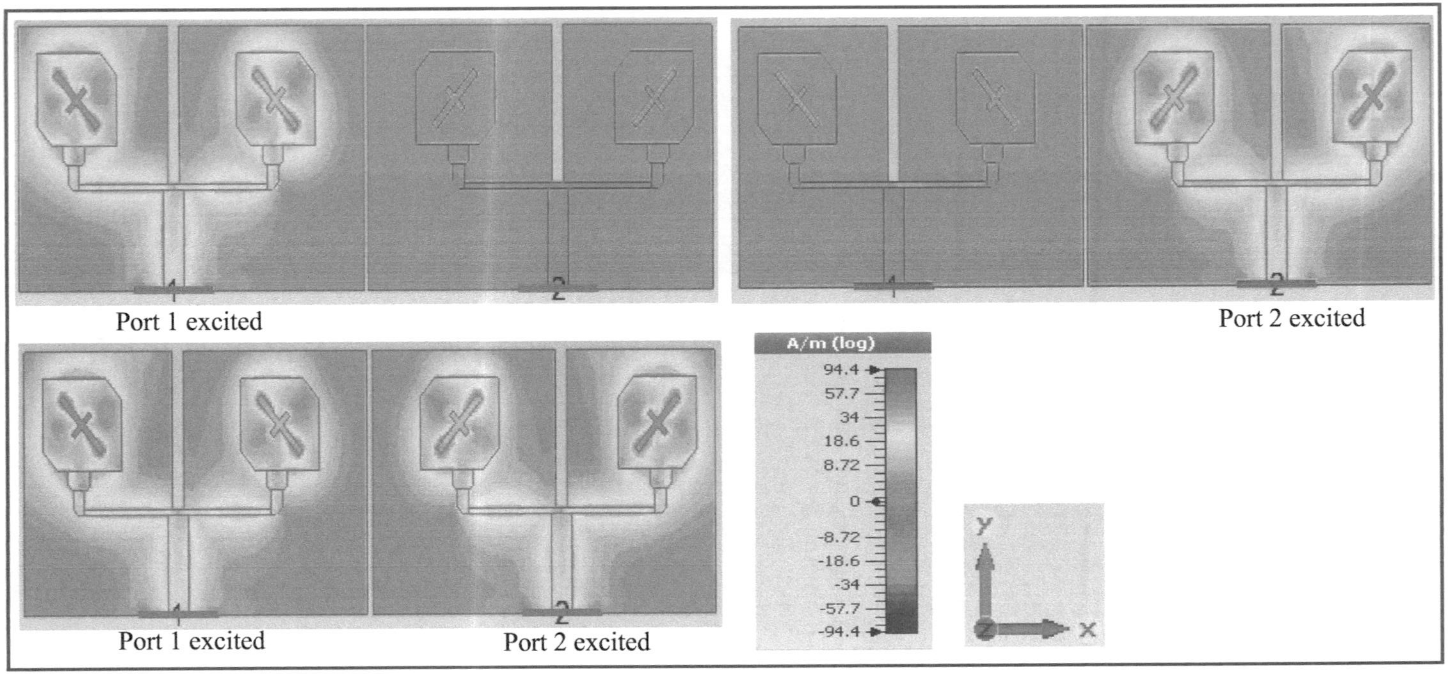

FIGURE 9.9
SCD of CP-MIMO antenna [228]. [From: Leeladhar Malviya, Rajib K. Panigrahi, and Machavaram V. Kartikeyan, "Circularly polarized 2×2 MIMO antenna for WLAN applications," Progress in Electromagnetics Research C, vol. 66, pp. 97–107, July 2016. Reproduced courtesy of the Electromagnetics Academy.]

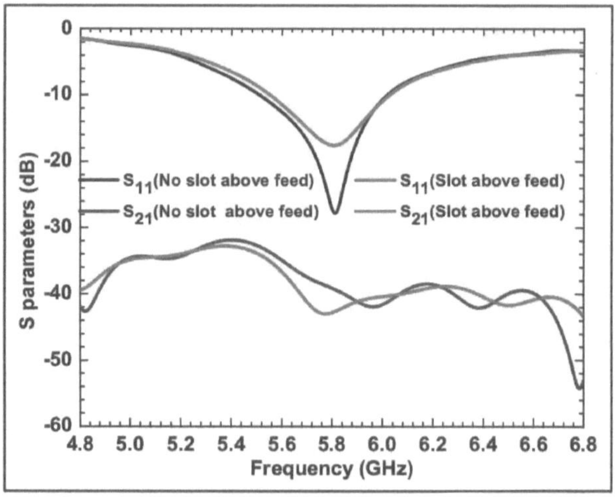

FIGURE 9.10
S parameters with and without the slot above feed line [228]. [From: Leeladhar Malviya, Rajib K. Panigrahi, and Machavaram V. Kartikeyan, "Circularly polarized 2 × 2 MIMO antenna for WLAN applications," Progress in Electromagnetics Research C, vol. 66, pp. 97–107, July 2016. Reproduced courtesy of the Electromagnetics Academy.]

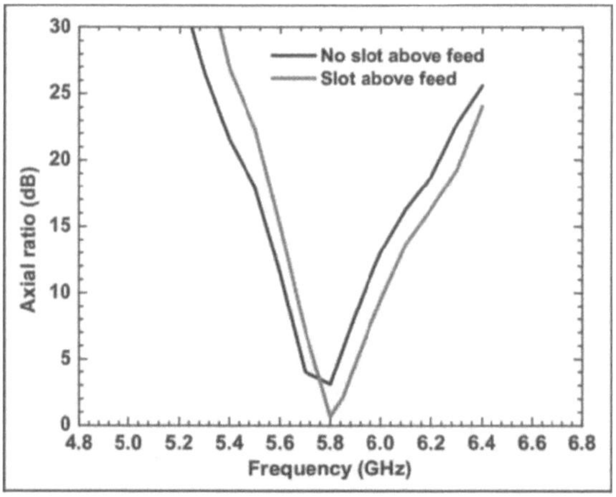

FIGURE 9.11
Axial ratio with and without the slot above feed line [228]. [From: Leeladhar Malviya, Rajib K. Panigrahi, and Machavaram V. Kartikeyan, "Circularly polarized 2 × 2 MIMO antenna for WLAN applications," Progress in Electromagnetics Research C, vol. 66, pp. 97–107, July 2016. Reproduced courtesy of the Electromagnetics Academy.]

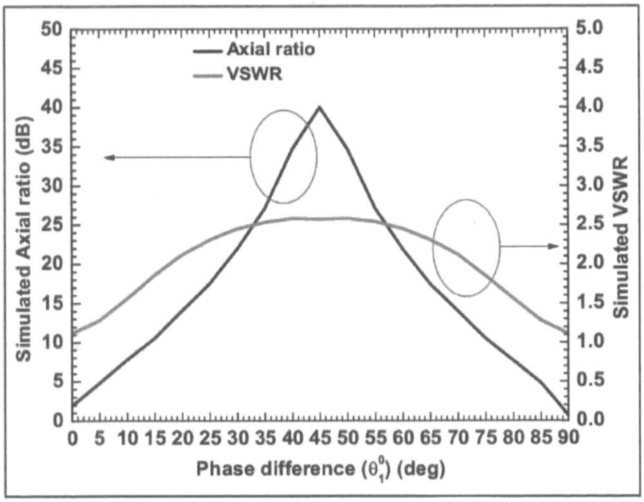

FIGURE 9.12
Effect of phase difference of asymmetric slots on CP-MIMO [228]. [From: Leeladhar Malviya, Rajib K. Panigrahi, and Machavaram V. Kartikeyan, "Circularly polarized 2 × 2 MIMO antenna for WLAN applications," Progress in Electromagnetics Research C, vol. 66, pp. 97–107, July 2016. Reproduced courtesy of the Electromagnetics Academy.]

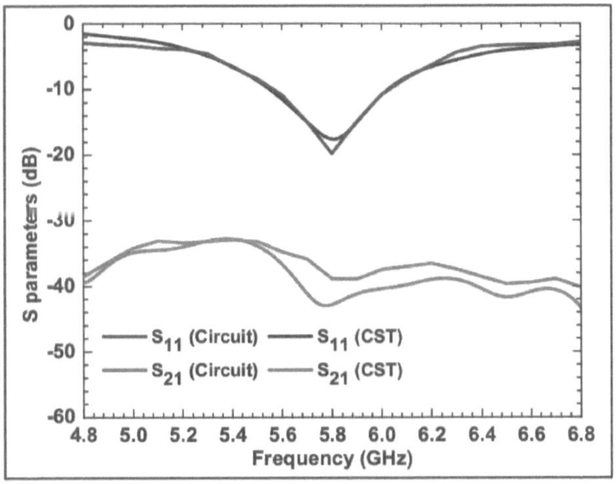

FIGURE 9.13
Equivalent circuit and CST *S* parameters [228]. [From: Leeladhar Malviya, Rajib K. Panigrahi, and Machavaram V. Kartikeyan, "Circularly polarized 2 × 2 MIMO antenna for WLAN applications," Progress in Electromagnetics Research C, vol. 66, pp. 97–107, July 2016. Reproduced courtesy of the Electromagnetics Academy.]

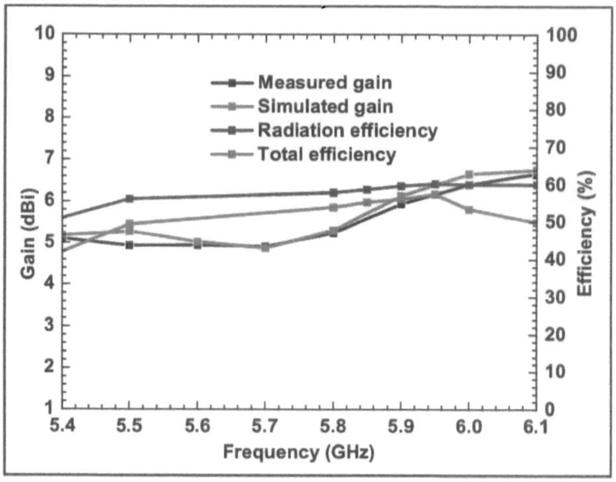

FIGURE 9.14
Gain and efficiency of CP-MIMO antenna [228]. [From: Leeladhar Malviya, Rajib K. Panigrahi, and Machavaram V. Kartikeyan, "Circularly polarized 2 × 2 MIMO antenna for WLAN applications," Progress in Electromagnetics Research C, vol. 66, pp. 97–107, July 2016. Reproduced courtesy of the Electromagnetics Academy.]

measured and simulated values is due to fabrication errors and port/cable coupling losses.

TARC of the CP-MIMO antenna is obtained by (4.10). Figure 9.18 shows the measured TARC with the random nature of the signals and their excitation angles at different ports. Obtained curves have a resemblance with isolation parameter S_{21}. As per CP characteristics, the closeness of curves shows the effectiveness with all the excitation angles, i.e. CP antenna works for all weather conditions. CP-MIMO has active bandwidth of 500 MHz for −20 dB return-loss bandwidth.

MEG of the CP-MIMO is obtained using (4.15). Let the Gaussian/Uniform medium signals have mean $(\mu) = 0$ and variance $(\sigma) = 20$, for both horizontal and vertical components. Figure 9.19 shows the isotropic medium MEG with $XPR = 0$ dB as −3.5 dB and is almost constant in the whole band, and for $XPR = 6$ dB, MEG lies in the range of −1.6 to −3.5 dB. Similarly, MEG for Gaussian medium with $XPR = 0$ dB, lies in the range of −4.3 to −6.5 dB, and for $XPR = 6$ dB, MEG lies in the range of −2.7 to −4.9 dB. The simulated values of MEGs for 5.8 GHz resonant frequency for Isotropic medium for $XPR = 0$ dB and $XPR = 6$ dB are less than −2.25 dB, and for Gaussian medium, these values are less than −6.46 dB. Hence, CP-MIMO antenna is suitable for indoor and outdoor Gaussian environments, and indoor Isotropic environment.

Measurement of far-field radiation patterns of the CP-MIMO antenna is performed in an anechoic chamber at 5.8 GHz resonant frequency for E-field and H-field validation at each port. Unused port in an anechoic chamber is terminated with 50.0 Ω to avoid any noise pick-up. CP-MIMO antenna elements

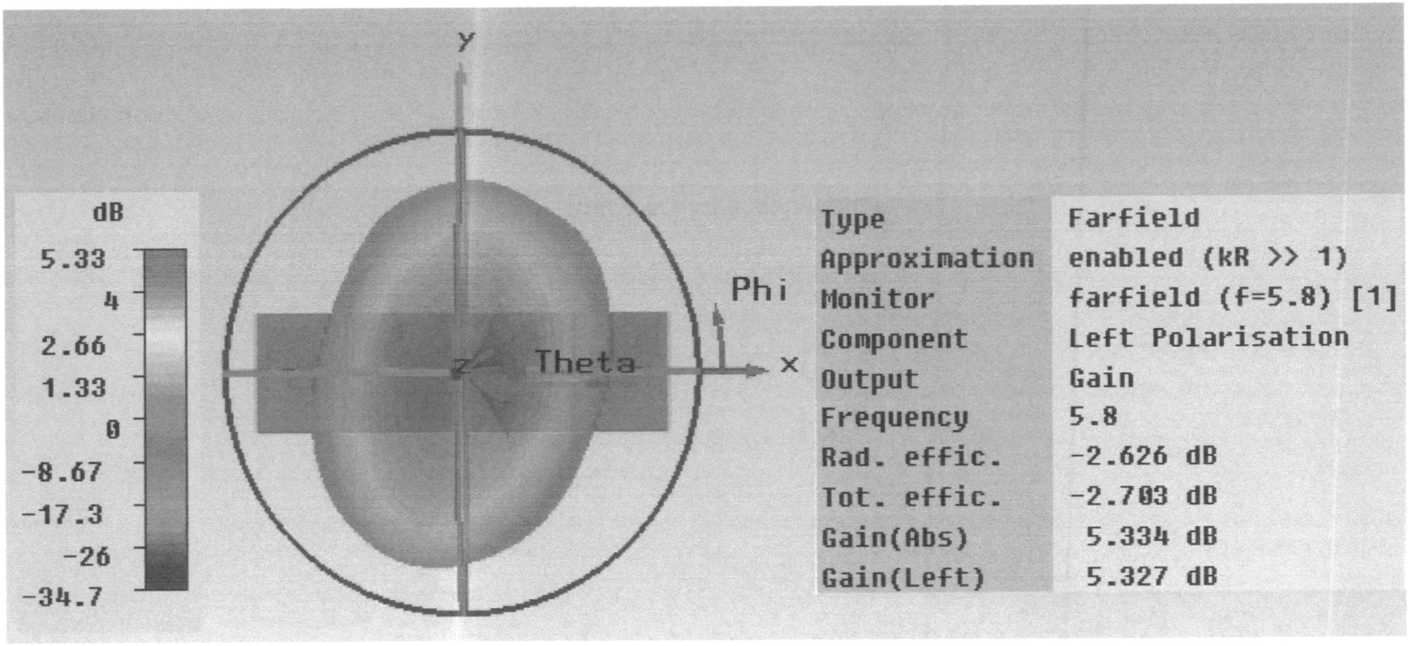

FIGURE 9.15
Left hand polarization pattern at 5.8 GHz [228]. [From: Leeladhar Malviya, Rajib K. Panigrahi, and Machavaram V. Kartikeyan, "Circularly polarized 2 × 2 MIMO antenna for WLAN applications," Progress in Electromagnetics Research C, vol. 66, pp. 97–107, July 2016. Reproduced courtesy of the Electromagnetics Academy.]

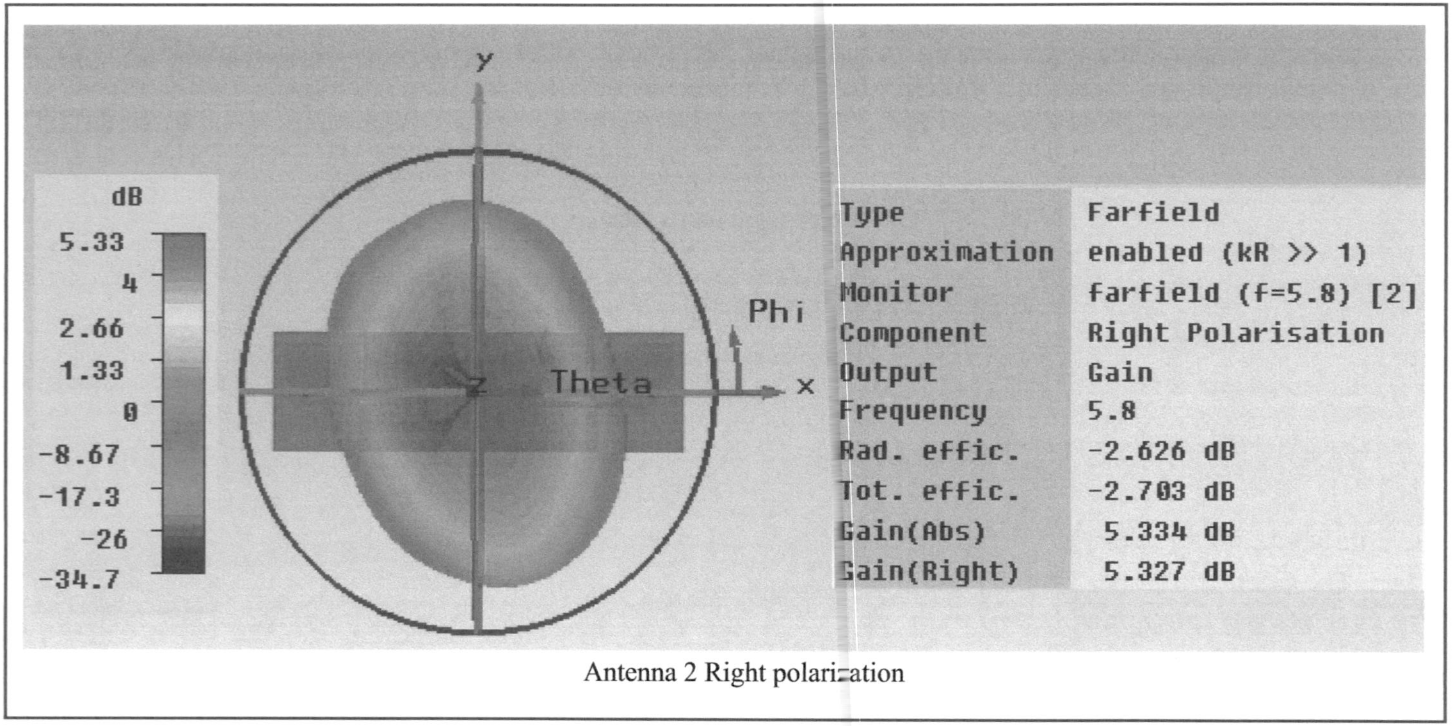

FIGURE 9.16
Right hand polarization pattern at 5.8 GHz [228]. [From: Leeladhar Malviya, Rajib K. Panigrahi, and Machavaram V. Kartikeyan, "Circularly polarized 2 × 2 MIMO antenna for WLAN applications," Progress in Electromagnetics Research C, vol. 66, pp. 97–107, July 2016. Reproduced courtesy of the Electromagnetics Academy.]

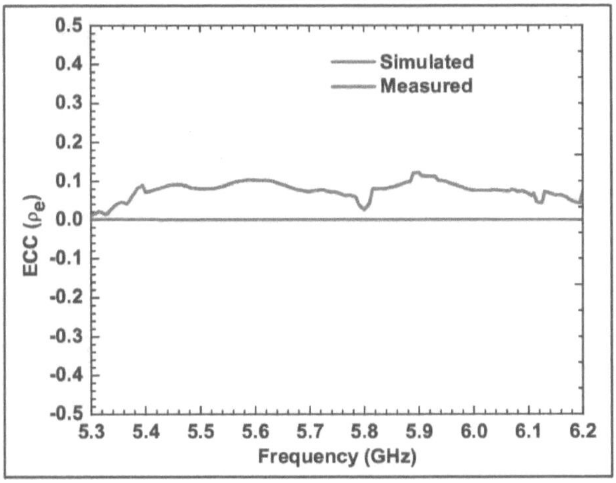

FIGURE 9.17
ECC of CP-MIMO antenna [228]. [From: Leeladhar Malviya, Rajib K. Panigrahi, and Machavaram V. Kartikeyan, "Circularly polarized 2 × 2 MIMO antenna for WLAN applications," Progress in Electromagnetics Research C, vol. 66, pp. 97–107, July 2016. Reproduced courtesy of the Electromagnetics Academy.]

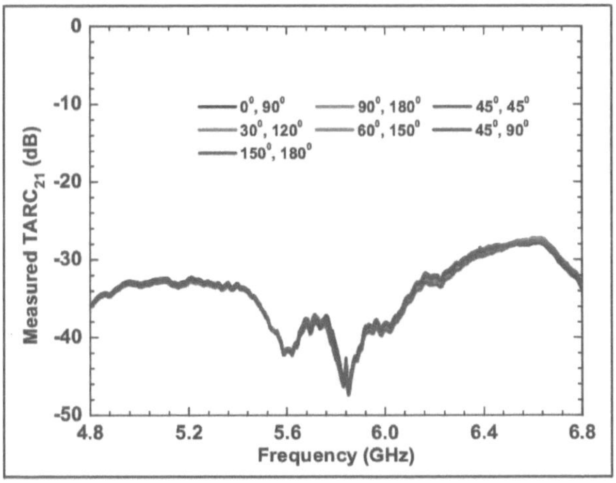

FIGURE 9.18
TARC of CP-MIMO antenna [228]. [From: Leeladhar Malviya, Rajib K. Panigrahi, and Machavaram V. Kartikeyan, "Circularly polarized 2 × 2 MIMO antenna for WLAN applications," Progress in Electromagnetics Research C, vol. 66, pp. 97–107, July 2016. Reproduced courtesy of the Electromagnetics Academy.]

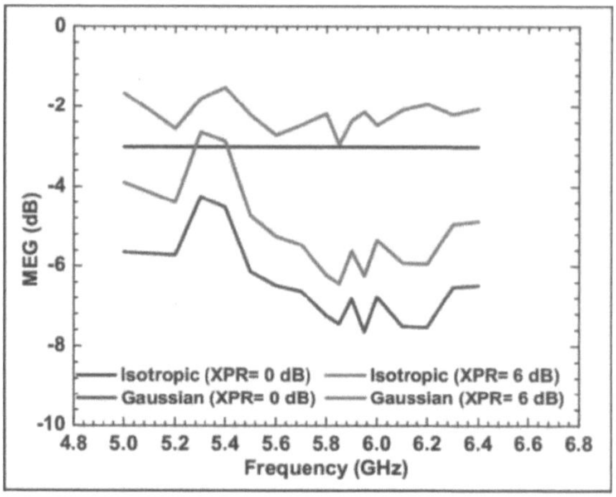

FIGURE 9.19
MEG of CP-MIMO antenna [228]. [From: Leeladhar Malviya, Rajib K. Panigrahi, and Machavaram V. Kartikeyan, "Circularly polarized 2 × 2 MIMO antenna for WLAN applications," Progress in Electromagnetics Research C, vol. 66, pp. 97–107, July 2016. Reproduced courtesy of the Electromagnetics Academy.]

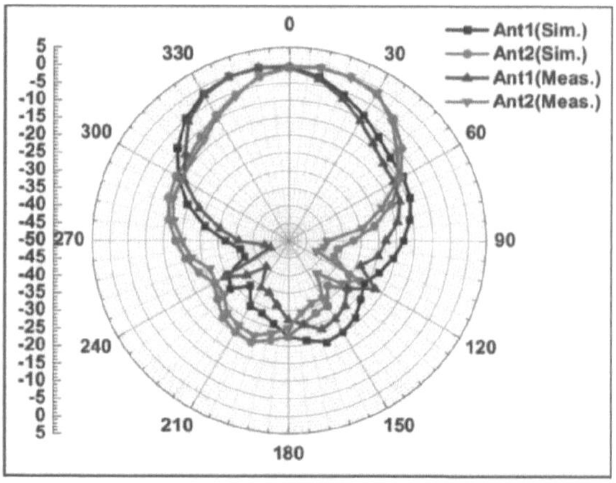

FIGURE 9.20
E-field radiation patterns [228]. [From: Leeladhar Malviya, Rajib K. Panigrahi, and Machavaram V. Kartikeyan, "Circularly polarized 2 × 2 MIMO antenna for WLAN applications," Progress in Electromagnetics Research C, vol. 66, pp. 97–107, July 2016. Reproduced courtesy of the Electromagnetics Academy.]

with the mirrored position, exhibit balanced radiation patterns. Figure 9.20 shows, simulated and measured E-field radiation patterns, where, main lobe directions are ±10°, and the angular widths are 57.6° at each port. Figure 9.21

shows, H-field radiation patterns, where main lobe directions are ±10°, and the angular widths are 92.5°. Due to 1 × 2 feed structure, these patterns are directional.

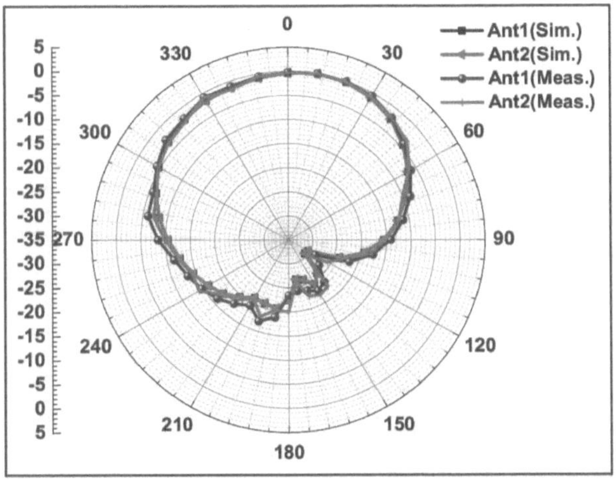

FIGURE 9.21
H-field radiation patterns [228]. [From: Leeladhar Malviya, Rajib K. Panigrahi, and Machavaram V. Kartikeyan, "Circularly polarized 2 × 2 MIMO antenna for WLAN applications," Progress in Electromagnetics Research C, vol. 66, pp. 97–107, July 2016. Reproduced courtesy of the Electromagnetics Academy.]

9.4 Concluding Remarks

CP-MIMO antenna with two 1 × 2 feed arms resonated at 5.8 GHz for IEEE 802.11 standard. Two asymmetric slots with 90° phase difference have played the major role to make it circularly polarized. Various aspects of CP-MIMO antenna have been investigated in this chapter for the better characterization of S parameters and axial ratio. CP-MIMO antenna exhibited 5.49–6.02 GHz frequency band and greater than 33.0 dB isolation between ports. Achieved ECC is very much less than the limit set by the ITU for wireless communication. Designed CP-MIMO is capable in all weather conditions, as justified by the TARC excitation angles. From MEG values, it shows the suitability for indoor and outdoor Gaussian environments completely [228].

10

MIMO Antenna Designs with Diversity Techniques for LTE Applications

The long-term evolution (LTE) is the fastest growing and inter-operable wireless technology of third generation partnership project (3GPP) for the fourth generation and contributes to the next generations. High signal-to-noise ratio (SNR), low latency, and high data rates are the contributions of LTE in conjunction with MIMO antennas. LTE has backward compatibility and scalable bandwidth for uplink and downlink channels. LTE without the MIMO is equivalent to a dream that cannot come true because voice/video/multimedia on-demand services are the joint ventures of these two only. This chapter presents two LTE designs for multipath propagation (or NLOS communication). First, MIMO design is based on the C-shaped folded loop radiator, and the second, MIMO design is based on a mathematically inspired shape.

10.1 Introduction

MIMO with LTE is a solution to the fastest growing population of wireless users. Variety of wireless applications in indoor and outdoor activities require scalable bandwidth and variable data rates to cooperate with all kinds of users. In this chapter two LTE designs are presented to accommodate certain wireless bands. First design uses a C-shaped folded loop structure to form 2×2 MIMO antenna. In the second design, a mathematically inspired patch shape is used to form 2×2 MIMO antenna. Both the design covers 1.8/1.9 GHz wireless applications and have wide-band responses with respect to -10 dB return-loss band. All the design parameters of both the presented MIMO are thoroughly evaluated and measured with VNA and in an anechoic chamber.

10.2 Design and Analysis of C-Shaped Folded Loop MIMO Radiator

MIMO antenna with C-shaped folded loop radiator is presented for the coverage of LTE band in this section.

10.2.1 Introduction and Related Work

4G-LTE MIMO antennas are designed to provide scalable bandwidth and variable data rates required for various indoor and outdoor activities. MIMO antenna designers are working on the user-friendly technologies with whole frequency utilization, lowest power requirements, proper link adaptation/utilization, and multi-channel propagation issues. MIMO satisfies the demand of high data rates, though compactness with better performance parameters is the prime requirement for all the modern wireless equipment/ gadgets.

Different mutual coupling reduction techniques and diversity techniques lead to the compactness and better performance characterization of designed MIMO antennas for wireless needs. Miniaturization of multi-element antenna is a compromising challenge, and needs the participation of all the radiating elements for desired/targeted diversity and far-field characterization [229,230].

Variety of 1×1 and 2×1 radiators were designed for single and multi-port wireless applications like shorted antenna layers [231], flip in element for band switching [232], slots for size reduction [233], capacitive strips [234], reconfigurable slots [235], and fractal geometries [236]. Similarly, labyrinth split ring resonator (LSRR) with meaner lines was used to improve isolation characteristics [237]. Open/Short slots provide compactness, impedance matching, and independent tuning in MIMO antennas [238, 239].

Most of the four-port MIMO radiators were designed to cover IEEE 802.11 WLAN (2.4/5.2/5.8) bands and IEEE 802.16 WiMAX (2.5/3.5/5.5) bands. First design consists of C-shaped folded loop MIMO with more than 10 dB isolation among different ports.

10.2.2 Folded Loop MIMO Antenna Design and Implementation

Four-port LTE MIMO with C-shaped folded loop radiators is designed for compactness and enhanced performance characteristics. All the radiating elements with 50.0 Ω ports and diversity arrangement are utilized to transmit and receive operations. The designed shape results in longer current paths and is a continuous monopole loop. Thus, achieving the compactness of radiator for 1.65–2.17 GHz frequency range. FR-4 dielectric substrate (permittivity of 4.4, loss tangent of 0.025, and thickness of 1.524 mm) of dimension 68 × 96 mm^2 has been chosen to design the folded loop MIMO with CST-MWS simulation tool.

The frequency of operation = 1.6-2.1 GHz, VSWR = 2:1, isolation >10 dB, gain (dBi) >5, efficiency (%) >90, ECC <0.1, and LTE application are the set goals of the MIMO antenna design here. Figures 10.1 and 10.2 show the single element and fabricated folded loop MIMO antenna. The optimized parameters are given in Table 10.1.

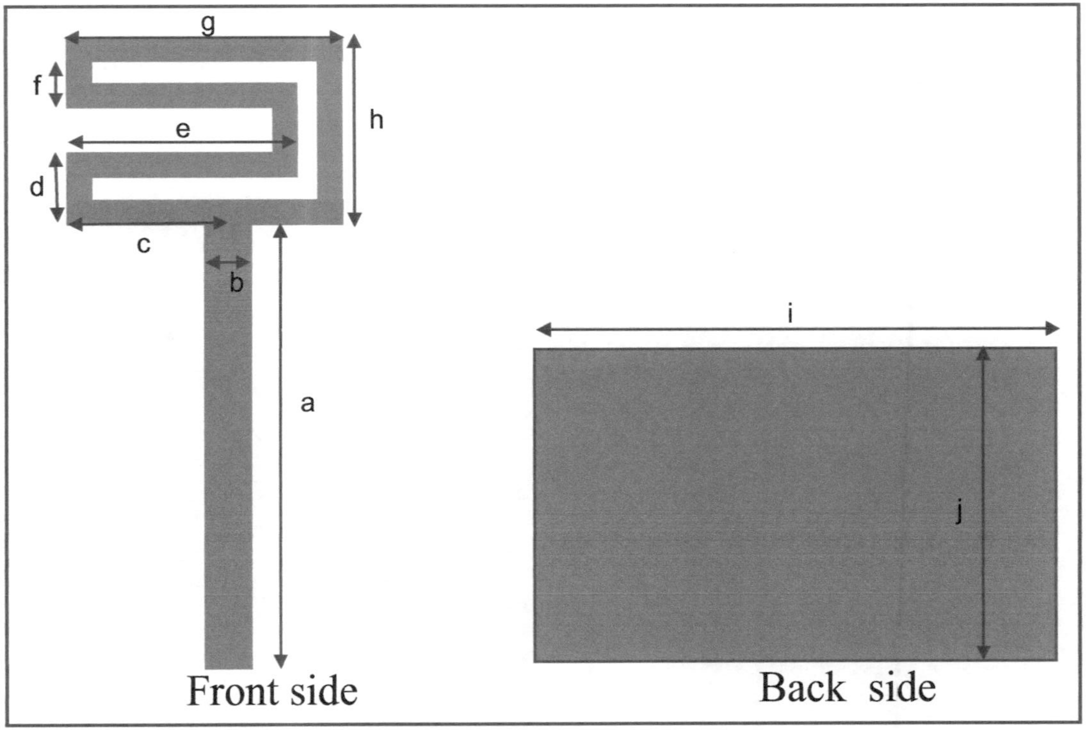

FIGURE 10.1
Schematic views of single folded loop [240]. [From: Leeladhar Malviya, Rajib K. Panigrahi, and M. V. Kartikeyan, "Four element planar MIMO antenna design for long term evolution operation," IETE Journal of Research, vol. 64, no. 3, pp. 367–373, August 2017. Reproduced courtesy of the Taylor and Francis.]

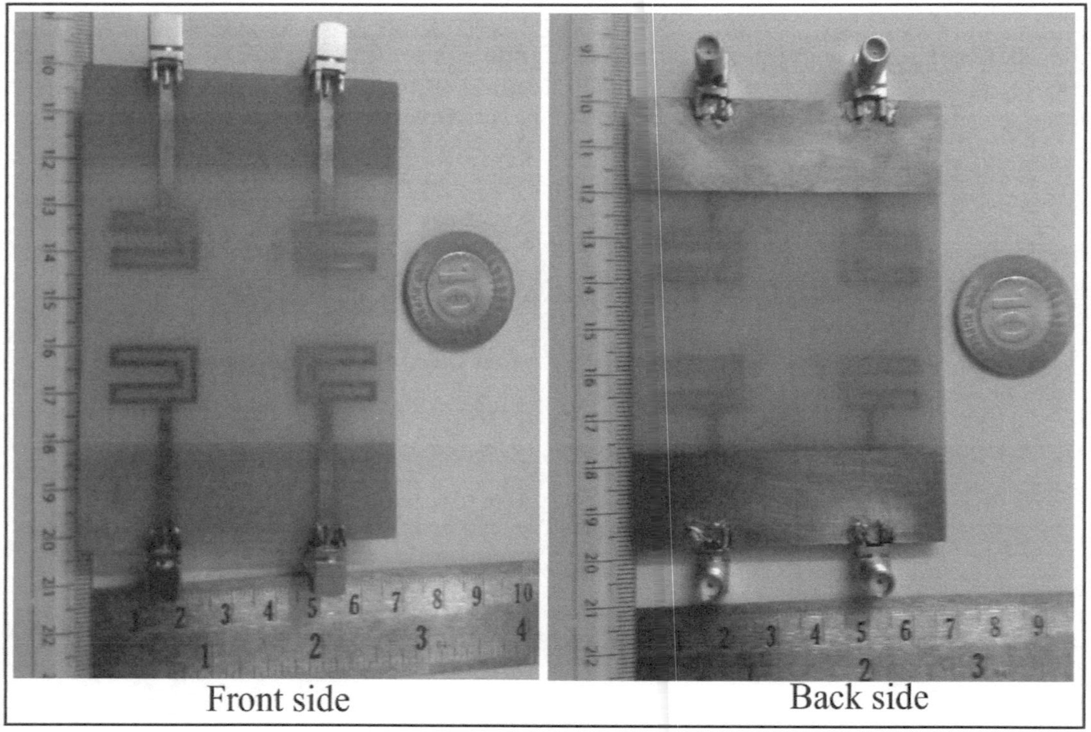

FIGURE 10.2
Fabricated views of folded loop MIMO antenna [240]. [From: Leeladhar Malviya, Rajib K. Panigrahi, and M. V. Kartikeyan, "Four element planar MIMO antenna design for long term evolution operation," IETE Journal of Research, vol. 64, no. 3, pp. 367–373, August 2017. Reproduced courtesy of the Taylor and Francis.]

TABLE 10.1
Critical parameters of folded loop MIMO (Unit: mm) [240]. [From: Leeladhar Malviya, Rajib K. Panigrahi, and M. V. Kartikeyan, "Four element planar MIMO antenna design for long term evolution operation," IETE Journal of Research, vol. 64, no. 3, pp. 367–373, August 2017. Reproduced courtesy of the Taylor and Francis.]

Parameter	a	b	c	d	e
Value	28.5	2.96	11.38	4.44	14.92
Parameter	f	g	h	i	j
Value	2.96	17.88	12.0	34.0	20.0

Figure 10.3 shows the equivalent circuit analysis of the folded loop MIMO antenna using ADS software. Folded loop MIMO antenna resonant peak (S_{11}) is represented by parallel combination of L1–C1 components. Similarly, isolation parameter S_{12} by parallel combination of L2–C2, S_{13} by L3–C3, and S_{14} by L4–C4. The values of circuit components are: L1 = 4.20 nH, C1 = 0.18 pf, L2 = 2.38 nH, C2 = 0.6 pf, L3 = 1.8 nH, C3 = 0.28 pf, L4 = 0.96 nH, and C4 = 0.28 pf.

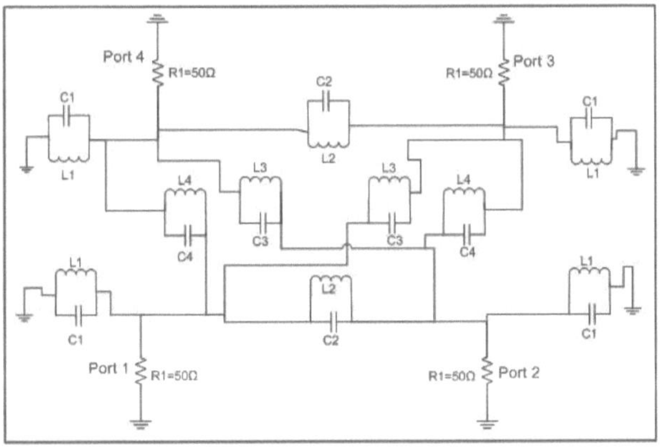

FIGURE 10.3
Equivalent circuit model of folded loop MIMO antenna [240]. [From: Leeladhar Malviya, Rajib K. Panigrahi, and M. V. Kartikeyan, "Four element planar MIMO antenna design for long term evolution operation," IETE Journal of Research, vol. 64, no. 3, pp. 367–373, August 2017. Reproduced courtesy of the Taylor and Francis.]

10.2.3 Simulation-Measurement Results and Discussion

Due to optimization of folded loop MIMO antenna, CST-MWS simulated and VNA measured S parameters are observed as, $S_{11} = S_{22} = S_{33} = S_{44}$,

$S_{12} = S_{21} = S_{34} = S_{43}$, $S_{13} = S_{31} = S_{24} = S_{42}$, and $S_{14} = S_{41} = S_{23} = S_{32}$. Such a situation is possible with 180° phase rotated and phase reversed radiating elements. For easy analysis of results S_{11}, S_{12}, S_{13}, and S_{14} scattering parameters are utilized only for this design.

Figure 10.4 shows the simulated and measured S parameters. It is observed that the designed folded loop MIMO has 2:1 VSWR band of 1.66–2.17 GHz (simulated), and 1.60–2.16 GHz (measured). Different isolations S_{12}, S_{13}, and S_{14} among the adjacent and diagonal ports are better than 10.0 dB in simulated and measured bands.

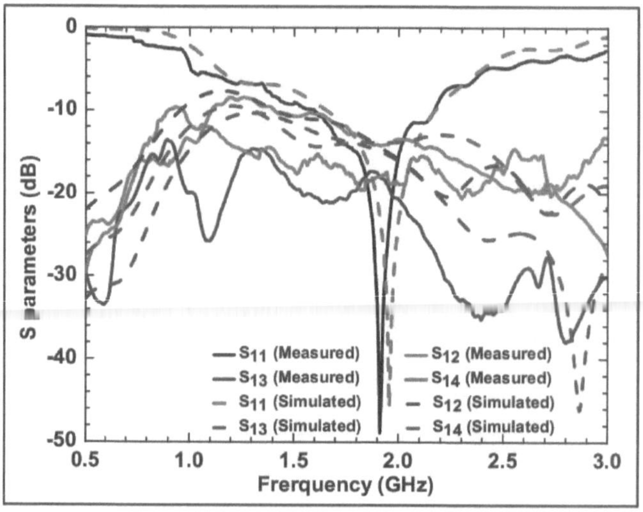

FIGURE 10.4
S parameters of folded loop MIMO antenna [240]. [From: Leeladhar Malviya, Rajib K. Panigrahi, and M. V. Kartikeyan, "Four element planar MIMO antenna design for long term evolution operation," IETE Journal of Research, vol. 64, no. 3, pp. 367–373, August 2017. Reproduced courtesy of the Taylor and Francis.]

Figure 10.5 shows the surface current distribution at 1.96 GHz resonant frequency under the excitation of one port (remaining ports are terminated by 50.0 Ω). Most of the current concentrates on feed arm. Some current links to the ports 1–2, and/or ports 3–4 due to common partial ground and is represented by S_{12} isolation parameter. Close proximity of folded loop MIMO radiators is also the reason for mutual coupling. Some current links due to 180° phase reversed radiators, corresponding isolations are counted using S_{13} and S_{14} parameters. These two reasons are used to calculate mutual coupling in presented folded loop MIMO. More than 10.0 dB of isolation is achieved among different radiators here. Surface current distribution at 1.96 GHz resonant frequency is in the 0–22.6 A/m range. Same concept is followed while exciting the other port or all the ports.

Figure 10.6 shows the effect of a conventional rectangular patch and selected folded loop. MIMO with conventional patch covers 2:1 VSWR band of 1.88–2.60 GHz frequency, and resonates at 2.23 GHz, with better than 15.0 dB of isolation at different ports. Presented folded loop MIMO covers 2:1 VSWR band of 1.66–2.17 GHz frequency with better than 11.0 dB in-band isolation among radiating ports. Folded-loop MIMO with the partial ground (full ground takes very big size as compared to partial ground for the same frequency) adds compactness in design. Placement of radiators have a high impact on maintaining the isolation.

Figure 10.7 shows the CST simulated and equivalent circuit responses. It is observed that S_{11}, S_{12}, S_{13}, and S_{14} parameters have very minute differences.

Far-field gain of the folded loop MIMO antenna is measured using the substitution method in the presence of standard horn antennas in the anechoic chamber. Figure 10.8 shows more than 2.5 dBi gain for both the simulated and measured frequency bands. At 1.96 GHz resonant frequency, the simulated MIMO gain is 2.8 dBi. Similarly, measured gain at 1.91 GHz resonant is 2.73 dBi. Simulated radiation efficiency is more than 96%, and total efficiency in the whole band is more than 71%.

ECC (ρ_{e12}, ρ_{e13}, and ρ_{e14}) for the folded loop MIMO antenna may be obtained by substituting the following terms: $S_{11} = S_{22} = S_{33} = S_{44}$, $S_{12} = S_{21} = S_{34} = S_{43}$, $S_{14} = S_{41} = S_{23} = S_{32}$, and $S_{13} = S_{31} = S_{24} = S_{42}$ in (4.7). Consider i = 1 to 2, j = 1 to 2, for any two elements, and N = 2, as two antenna elements are selected at a time. Figure 10.9 shows the simulated and measured ECC. It is observed that ECC is less than 0.3 for both the simulated and measured bands.

At 1.91 GHz resonant frequency, the simulated ECC values are $\rho_{e12} = 1.56 \times 10^{-6}$, $\rho_{e13} = 3.43 \times 10^{-6}$, and $\rho_{e14} = 1.87 \times 10^{-6}$. Similarly, the measured ECC values at 1.91 GHz frequency are $\rho_{e12} = 3.9 \times 10^{-2}$, $\rho_{e13} = 6.84 \times 10^{-4}$, and $\rho_{e14} = 7.85 \times 10^{-4}$. Similarly, At 1.96 GHz resonant frequency, the simulated ECC values are $\rho_{e12} = 1.43 \times 10^{-6}$, $\rho_{e13} = 2.86 \times 10^{-6}$, and $\rho_{e14} = 1.51 \times 10^{-6}$. Similarly, the measured ECC values at 1.96 GHz frequency are $\rho_{e12} = 1.3 \times 10^{-1}$, $\rho_{e13} = 1.8 \times 10^{-1}$, and $\rho_{e14} = 1.67 \times 10^{-4}$. Fabrication process and port/cable coupling losses lead to certain differences between simulated and measured values of ECC.

Diversity performance in terms of TARC of folded-loop MIMO is obtained using (4.10). Figure 10.10 shows the responses of TARC. Due to the random nature of signals, distortions are observed in measured values with the different combinations of excitation angles at ports 1–2, 1–3, and 1–4. For phase combination 90° and 180° at different ports, good results of TARC are obtained. TARC bandwidth is more than 500 MHz for the considered phase relation at different ports.

Another diversity parameter MEG is obtained using (4.15). Let the horizontal and vertical components of the Gaussian signals have mean (μ) = 0 and

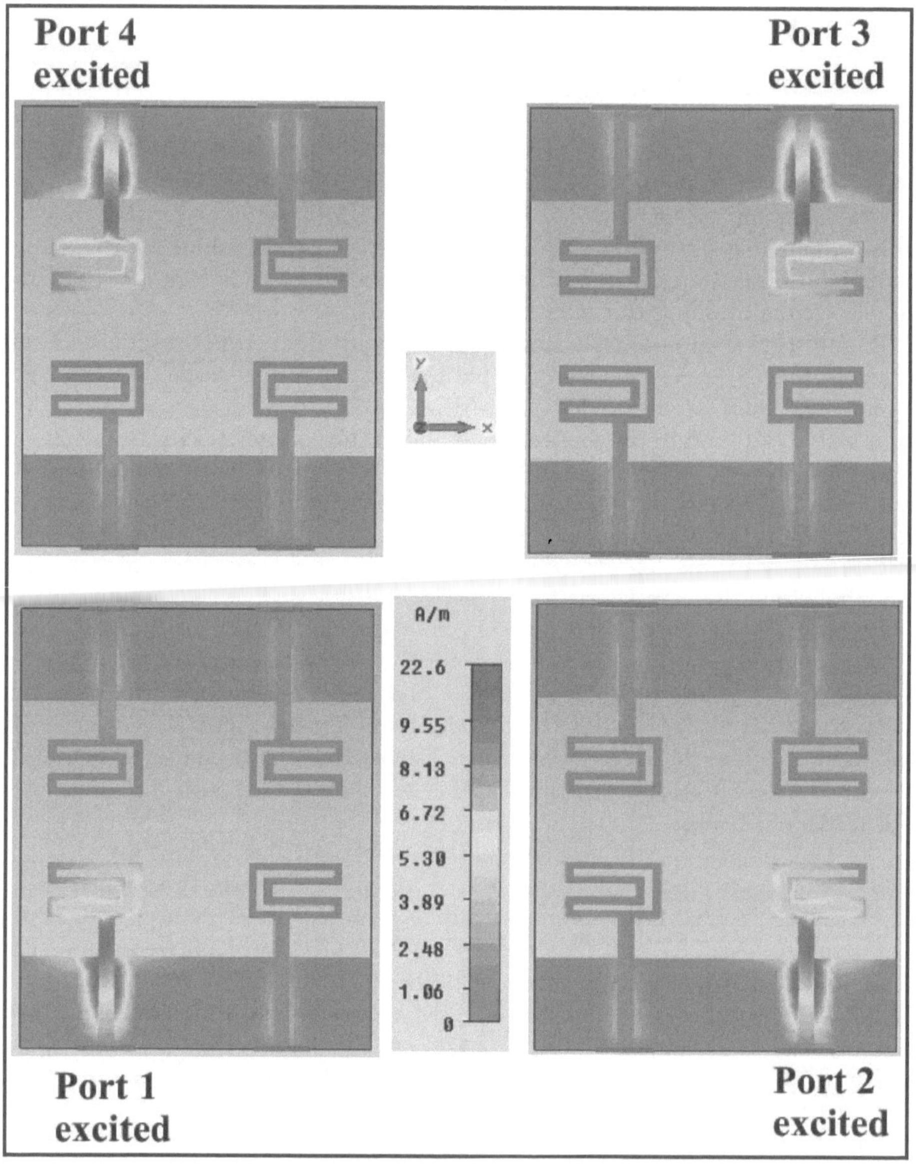

FIGURE 10.5
SCD of folded loop MIMO antenna at 1.96 GHz [240]. [From: Leeladhar Malviya, Rajib K. Panigrahi, and M. V. Kartikeyan, "Four element planar MIMO antenna design for long term evolution operation," IETE Journal of Research, vol. 64, no. 3, pp. 367–373, August 2017. Reproduced courtesy of the Taylor and Francis.]

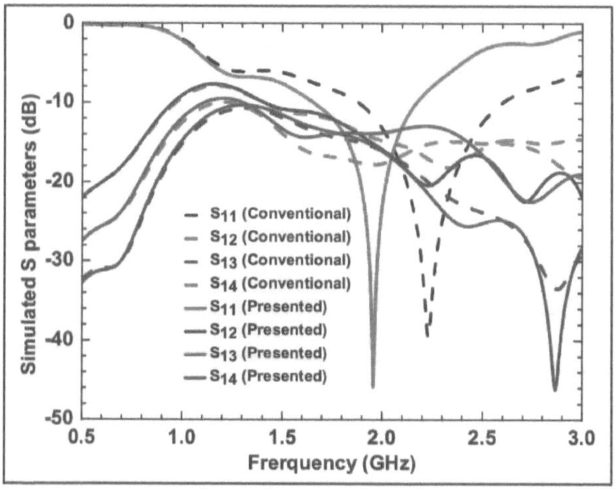

FIGURE 10.6
S parameters with conventional patch and folded loop MIMO [240]. [From: Leeladhar Malviya, Rajib K. Panigrahi, and M. V. Kartikeyan, "Four element planar MIMO antenna design for long term evolution operation," IETE Journal of Research, vol. 64, no. 3, pp. 367–373, August 2017. Reproduced courtesy of the Taylor and Francis.]

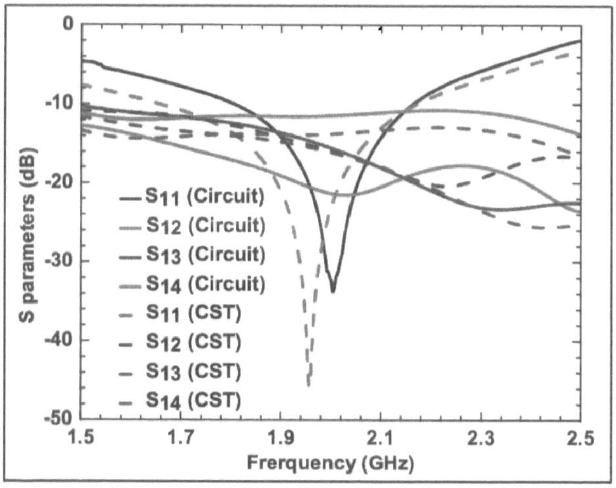

FIGURE 10.7
Equivalent circuit and CST S parameters [240]. [From: Leeladhar Malviya, Rajib K. Panigrahi, and M. V. Kartikeyan, "Four element planar MIMO antenna design for long term evolution operation," IETE Journal of Research, vol. 64, no. 3, pp. 367–373, August 2017. Reproduced courtesy of the Taylor and Francis.]

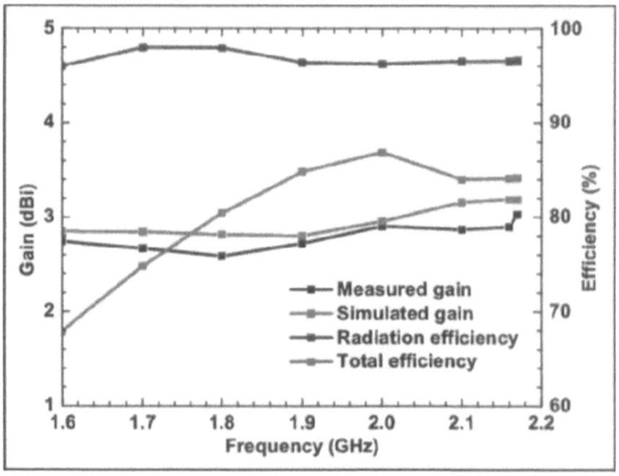

FIGURE 10.8
Gain and efficiency of folded loop MIMO antenna [240]. [From: Leeladhar Malviya, Rajib K. Panigrahi, and M. V. Kartikeyan, "Four element planar MIMO antenna design for long term evolution operation," IETE Journal of Research, vol. 64, no. 3, pp. 367–373, August 2017. Reproduced courtesy of the Taylor and Francis.]

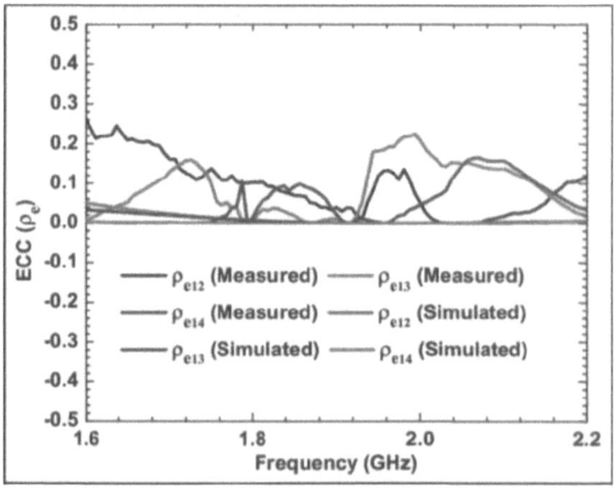

FIGURE 10.9
ECC of folded loop MIMO antenna [240]. [From: Leeladhar Malviya, Rajib K. Panigrahi, and M. V. Kartikeyan, "Four element planar MIMO antenna design for long term evolution operation," IETE Journal of Research, vol. 64, no. 3, pp. 367–373, August 2017. Reproduced courtesy of the Taylor and Francis.]

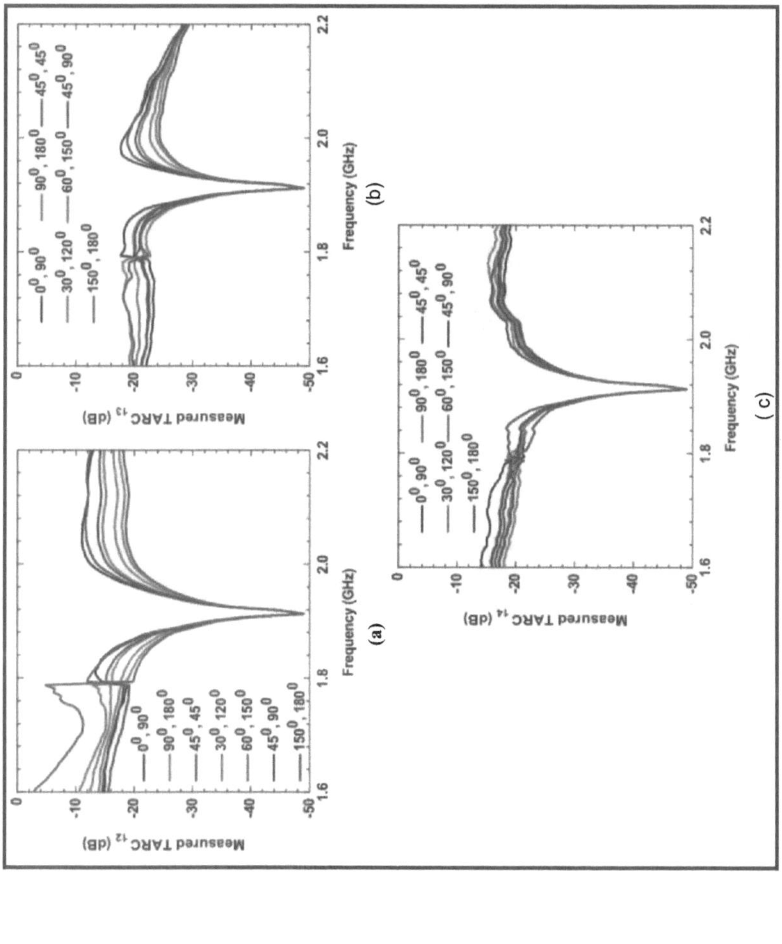

FIGURE 10.10
TARC of folded loop MIMO antenna [240]. [From: Leeladhar Malviya, Rajib K. Panigrahi, and M. V. Kartikeyan, "Four element planar MIMO antenna design for long term evolution operation," IETE Journal of Research, vol. 64, no. 3, pp. 367–373, August 2017. Reproduced courtesy of the Taylor and Francis.]

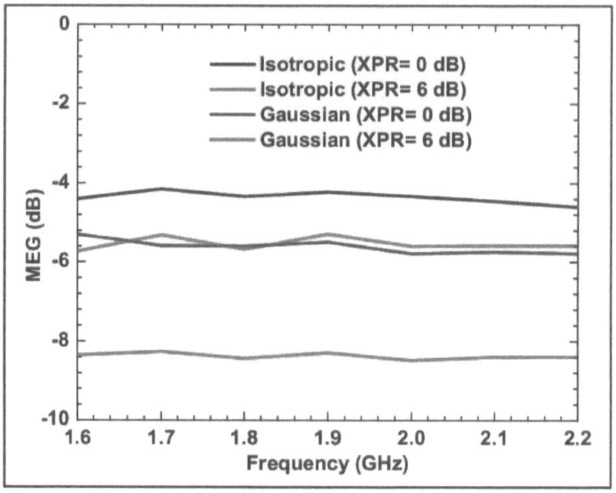

FIGURE 10.11
MEG of folded loop MIMO [240]. [From: Leeladhar Malviya, Rajib K. Panigrahi, and M. V. Kartikeyan, "Four element planar MIMO antenna design for long term evolution operation," IETE Journal of Research, vol. 64, no. 3, pp. 367–373, August 2017. Reproduced courtesy of the Taylor and Francis.]

variance (σ) = 20 respectively. Figure 10.11 shows different MEG responses for isotropic and Gaussian mediums. All the MEGs are considered with respect to $XPR = 0$ dB and $XPR = 6$ dB. Variation in MEGs is due to random variation in received powers with θ and ϕ components. Each MEG curve shows less than -4.0 dB in the whole frequency band. For simulated design, at 1.91 GHz resonant frequency MEG for an isotropic medium with $XPR = 0$ dB is -4.33 dB, and for $XPR = 6$ dB, MEG is -5.4 dB. Similarly, at 1.91 GHz frequency for Gaussian medium with $XPR = 0$ dB, MEG is -5.67 dB, and for $XPR = 6$ dB, MEG is -8.23 dB. For the fabricated MIMO antenna, at 1.96 GHz resonant frequency MEG for an isotropic medium with $XPR = 0$ dB is -4.54 dB, and for $XPR = 6$ dB, MEG is -5.51 dB. Similarly, at 1.96 GHz frequency for Gaussian medium with $XPR = 0$ dB, MEG is -5.73 dB, and for $XPR = 6$ dB, MEG is -8.31 dB. Hence, presented folded-loop MIMO design is completely useful is indoor and outdoor wireless LTE applications.

Far-field characterization of the folded loop MIMO is done in anechoic chamber. Figures 10.12 and 10.13 show the comparison of simulated and measured E-field patterns of folded loop MIMO at 1.91 GHz frequency. Similarly, Figures 10.14 and 10.15 show the simulated and measured H-field patterns of folded loop LTE MIMO antenna at 1.91 GHz resonant. Certain distortions in measured E-field and H-field patterns are observed due to the errors in the fabrication process, and port/cable coupling losses. Still, these are in good matching.

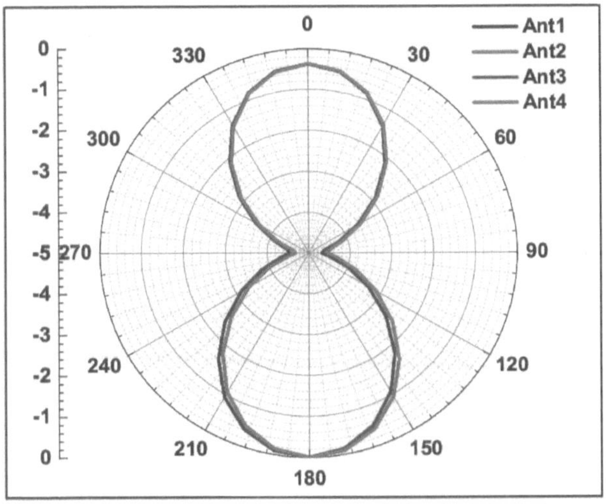

FIGURE 10.12
E-field patterns of folded loop MIMO at 1.91 GHz (simulated) [240]. [From: Leeladhar Malviya, Rajib K. Panigrahi, and M. V. Kartikeyan, "Four element planar MIMO antenna design for long term evolution operation," IETE Journal of Research, vol. 64, no. 3, pp. 367–373, August 2017. Reproduced courtesy of the Taylor and Francis.]

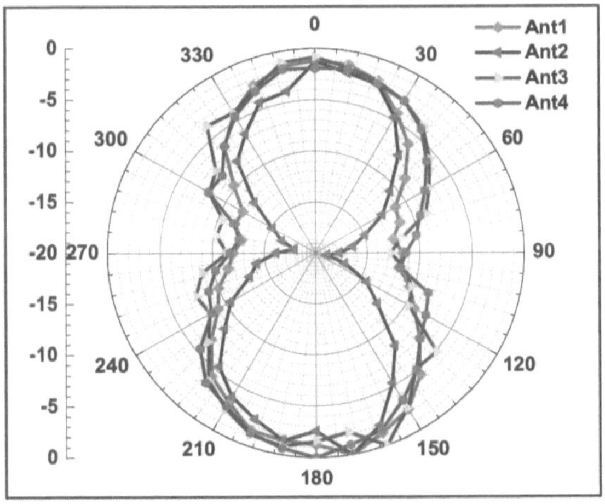

FIGURE 10.13
E-field patterns of folded loop MIMO at 1.91 GHz (measured) [240]. [From: Leeladhar Malviya, Rajib K. Panigrahi, and M. V. Kartikeyan, "Four element planar MIMO antenna design for long term evolution operation," IETE Journal of Research, vol. 64, no. 3, pp. 367–373, August 2017. Reproduced courtesy of the Taylor and Francis.]

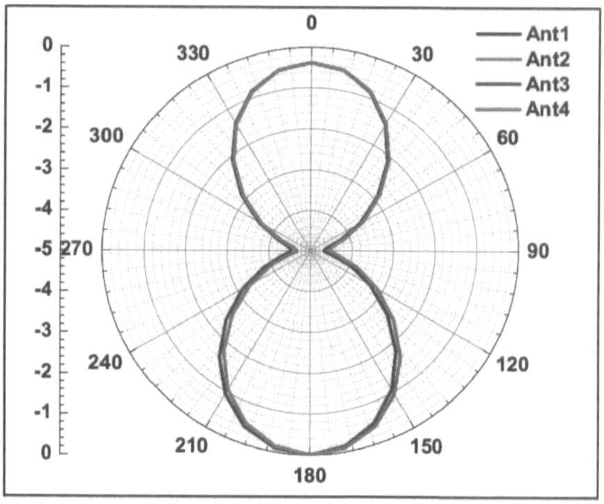

FIGURE 10.14
H-field patterns of folded loop MIMO at 1.91 GHz (simulated) [240]. [From: Leeladhar Malviya, Rajib K. Panigrahi, and M. V. Kartikeyan, "Four element planar MIMO antenna design for long term evolution operation," IETE Journal of Research, vol. 64, no. 3, pp. 367–373, August 2017. Reproduced courtesy of the Taylor and Francis.]

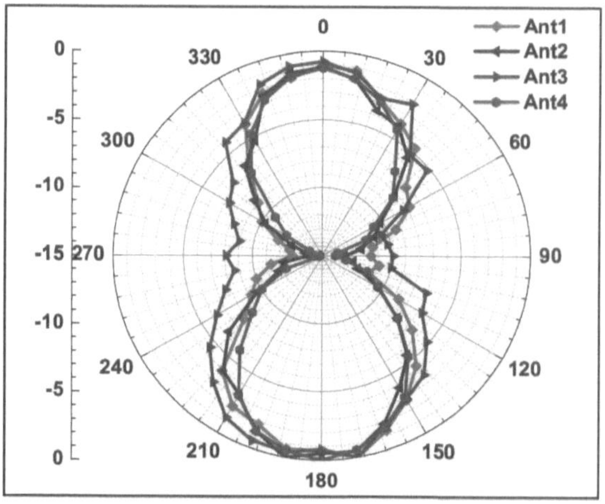

FIGURE 10.15
H-field patterns of folded loop MIMO at 1.91 GHz (measured) [240]. [From: Leeladhar Malviya, Rajib K. Panigrahi, and M. V. Kartikeyan, "Four element planar MIMO antenna design for long term evolution operation," IETE Journal of Research, vol. 64, no. 3, pp. 367–373, August 2017. Reproduced courtesy of the Taylor and Francis.]

A comparison of earlier reported works and presented design has been given in Table 10.2. The design given in reference [170] has a lower size for two elements only, and reference [171] has low size, but ECC is not disclosed and has a gain value of less than 1.0 dBi. The other designs have either high sizes and high ECC values. Therefore, the presented design provides better results in terms of size, gain, isolation, and ECC values in comparison with the earlier reported designs.

TABLE 10.2
Comparison of folded loop MIMO with earlier reported works [240] [From: Leeladhar Malviya, Rajib K. Panigrahi, and M. V. Kartikeyan, "Four element planar MIMO antenna design for long term evolution operation," IETE Journal of Research, vol. 64, no. 3, pp. 367–373, August 2017. Reproduced courtesy of the Taylor and Francis.]

Ref. No.	Frequency/ bands (GHz)	No. of elements	Size (mm^2)	Isolation (dB)	ECC	Gain (dBi)
[96]	1.8–2.9	4	140 × 120	>15	$<10^{-1}$	>5.5
[158]	1.96–2.36	4	200 × 200	>24	—	>4.0
[168]	1.71–2.69	2	118 × 58	>12	$<10^{-2}$	>1.0
[170]	1.71–1.88	2	80 × 50	>15	$<10^{-2}$	>2.18
[171]	1.88–2.20	4	95 × 60	>11.5	—	<1.0
In this design	1.65–2.17	4	68 × 96	>11	$<10^{-2}$	>2.7

10.3 Design and Analysis of Mathematically Inspired Dual Curved MIMO Radiator

Mathematical shapes like exponential curves, parabolic curves, and hyperbolic curves have an ability to exhibit variation in responses with respect to the square, cubic, and quadruple powers of the applied excitation signal. In this section, a mathamatically inspired MIMO for LTE band of operation is presented with polarization diversity approach.

10.3.1 Introduction and Related Work

Multipath fading, path loss, and shadowing are experienced in wireless and mobile devices. Fading and path loss are caused due to multiple reflection of signals from obstacles. Better spectral efficiency, solution of fading, gain, channel capacity, data rate, better QOS in NLOS environments are offered in modern MIMO antennas as compared to SISO antennas [241]. Current extension of MIMO technology can be seen in the form of stacking of MIMO antennas i.e. massive MIMO. Massive MIMO is an amazing application and great solution to the exponential growth of portable and wireless devices [242]. Wireless communication requires perfect channel models/coding, and fast signal processing approaches [243]. Base stations and mobile stations all are modernized with latest MIMO antennas for the better services to mobile and wireless users [244–246].

In this chapter, second design with mathematically inspired dual-curved four-port planar MIMO is presented with all the possible aspects. It covers 2:1 VSWR band of 1.68–2.24 GHz for LTE operation, and resonates at 1.9 GHz frequency. Feed line of the designed MIMO consists of a combination of 70.7 Ω and 50.0 Ω lines. More than 13.0 dB isolation, better than 3.0 dBi antenna gain, and 0.01 ECC value in the whole band are achieved here.

10.3.2 MIMO Antenna Design and Implementation

Mathematically inspired dual-curved MIMO is designed for 50.0 Ω ports using CST-MWS. FR-4 dielectric substrate of (thickness 1.524 mm, permittivity of 4.4, and loss tangent of 0.025) size 82.7 × 82.7 mm^2 is used for the MIMO prototype. A conventional rectangular patch is cut in curvature from both the sides to find the compact generalized radiating element. Four radiating elements with partial grounds are arranged in orthogonal polarization to achieve some size reduction, minimum return-loss, and sufficient isolation. Figures 10.16 and 10.17 show the schematic and fabricated MIMO prototypes. The optimized design parameters are given in Table 10.3.

TABLE 10.3
Optimized parameters of mathematically inspired MIMO (Unit: mm) [247]. [From: Leeladhar Malviya, Rajib K. Panigrahi, and M. V. Kartikeyan, "A low profile planar MIMO antenna with polarization diversity for LTE 1800/1900 applications," Microwave and Optical Technology Letters (MOTL), vol. 59, no. 3, pp. 533–538, March 2017. Reproduced courtesy of the John Wiley & Sons, Ltd.]

Parameter	a	b	c	d	e	f
Value	82.7	82.7	13.84	10.88	23.56	23.56
Parameter	g	h	i	j	k	r
Value	3.10	1.78	40.88	12.70	28.95	11.46

Figure 10.18 shows the equivalent circuit model of the mathematically inspired MIMO. Patch shape and combination of microstrip lines are represented by a parallel combination of inductor and capacitor components and each port by 50.0 Ω. Mathematically inspired patch with combined feed line is represented by a parallel combination of L1 and C1 components. Different isolations at ports 1–2, and ports 1–4 are represented by the parallel combination of L2–C2 and L4–C4. Mutual coupling at these ports is same, therefore, L2 = L4 and C2 = C4. Similarly, mutual coupling at ports 1–3 is represented by the parallel combination of L3–C3. Different values are given by: L1 = 0.68

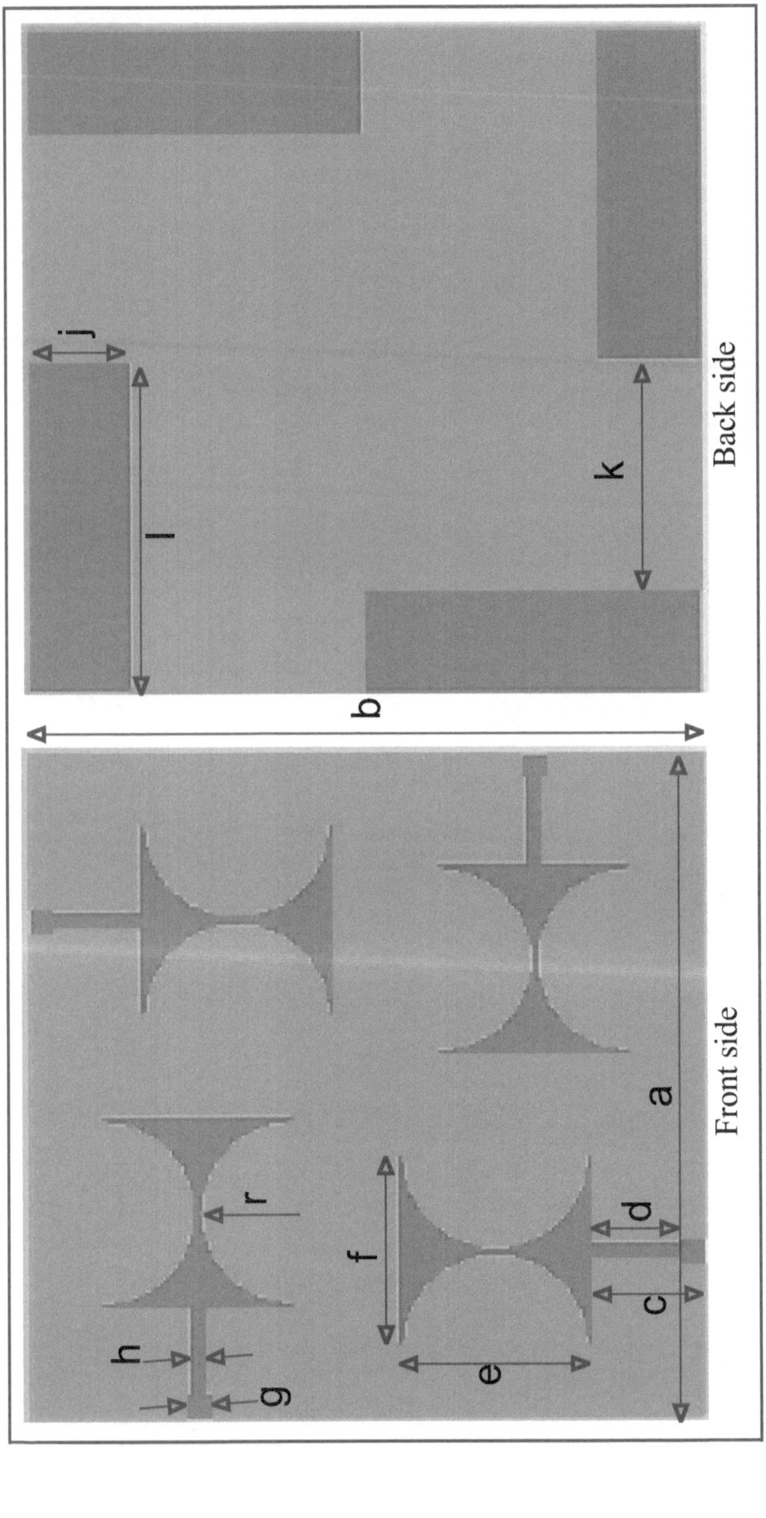

FIGURE 10.16
Schematic views of the mathematically inspired MIMO [247]. [From: Leeladhar Malviya, Rajib K. Panigrahi, and M. V. Kartikeyan, "A low profile planar MIMO antenna with polarization diversity for LTE 1800/1900 applications," Microwave and Optical Technology Letters (MOTL), vol. 59, no. 3, pp. 533–538, March 2017. Reproduced courtesy of the John Wiley & Sons, Ltd.]

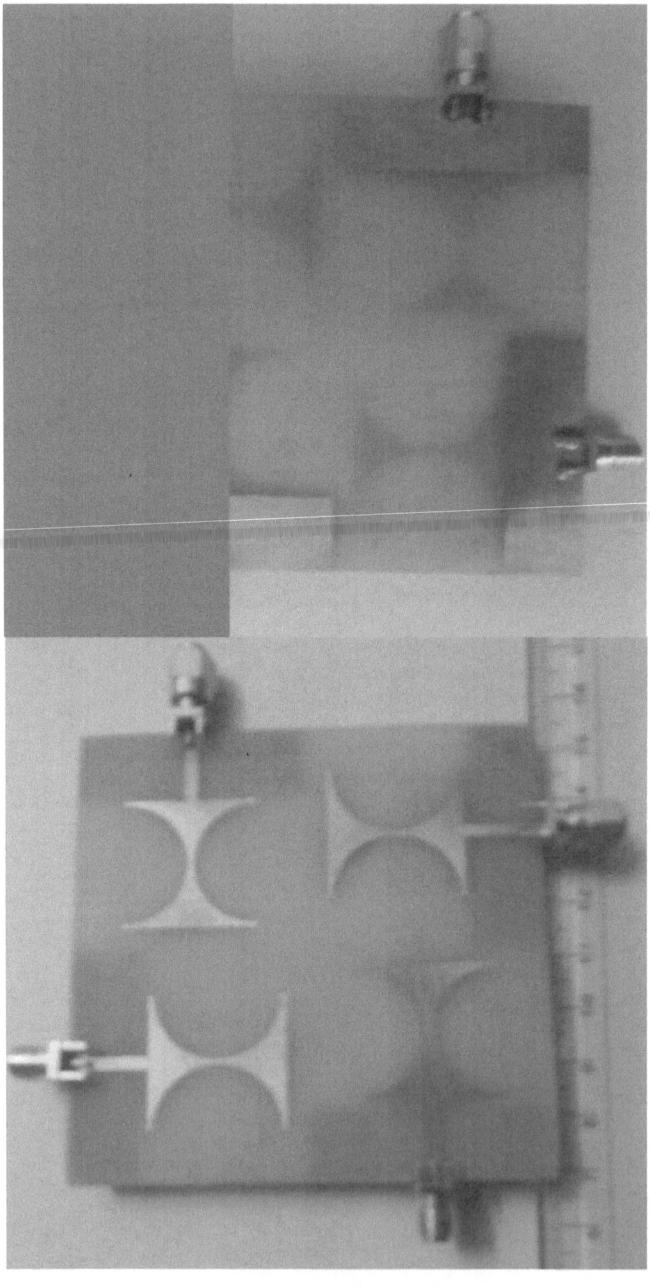

FIGURE 10.17
Fabricated views of mathematically inspired MIMO [247]. [From: Leelachar Malviya, Rajib K. Panigrahi, and M. V. Kartikeyan, “A low profile planar MIMO antenna with polarization diversity for LTE 1800/1900 applications,” Microwave and Optical Technology Letters (MOTL), vol. 59, no. 3, pp. 533–538, March 2017. Reproduced courtesy of the John Wiley & Sons, Ltd.]

nH, L2 = 1.8 nH, L3 = 1.53 nH, L4 = 1.8 nH, C1 = 2.86 pF, C2 = 0.5 pF, C3 = 0.47 pF, C4 = 0.5 pF, and R1 = 50.0 Ω.

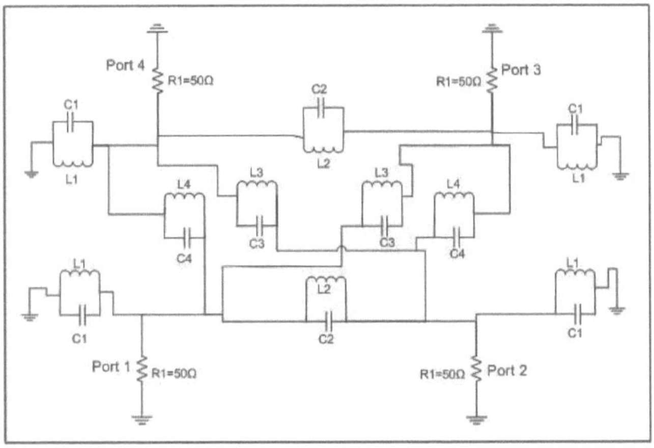

FIGURE 10.18
Equivalent circuit of mathematically inspired MIMO [247]. [From: Leeladhar Malviya, Rajib K. Panigrahi, and M. V. Kartikeyan, "A low profile planar MIMO antenna with polarization diversity for LTE 1800/1900 applications," Microwave and Optical Technology Letters (MOTL), vol. 59, no. 3, pp. 533–538, March 2017. Reproduced courtesy of the John Wiley & Sons, Ltd.]

10.3.3 Simulation-Measurement Results and Discussion

Mathematically inspired MIMO is a generalized structure which solves the drawbacks of a single element. The design also limits the surface waves due to orthogonal polarization. Due to placement of radiators, it is observed that return-loss parameters $S_{11} = S_{22} = S_{33} = S_{44}$, and isolation parameters $S_{12} = S_{21} = S_{14} = S_{41} = S_{32} = S_{23} = S_{34} = S_{43}$, and at diagonal ports $S_{13} = S_{31} = S_{42} = S_{24}$. Return-loss at ports is same because of the transformation of a single element, and isolations at different ports are dependent on space between them. For easy analysis of results, consider only S_{11}, S_{12}, and S_{13} scattering parameters. Due to equi-distance from port 1, $S_{12} = S_{14}$.

The conventional rectangular patch-based MIMO and the designed mathematically inspired MIMO are discussed here for the effectiveness of four-port design. Figure 10.19 shows that the conventional patch-based MIMO with partial ground covers 2:1 VSWR band of 2.98–3.34 GHz. MIMO antenna resonates at 3.16 GHz frequency, where the return-loss is −15.12 dB. Also, the bandwidth is 360.0 MHz. Port to port isolations at different ports are (S_{12}, S_{13}, and S_{14}) better than 14.0 dB. Mathematically inspired dual-curved patch-based MIMO covers 2:1 VSWR band of 1.68–2.24 GHz. Here, MIMO resonates at 1.9 GHz frequency, where the return-loss is −29.36 dB. The bandwidth, in

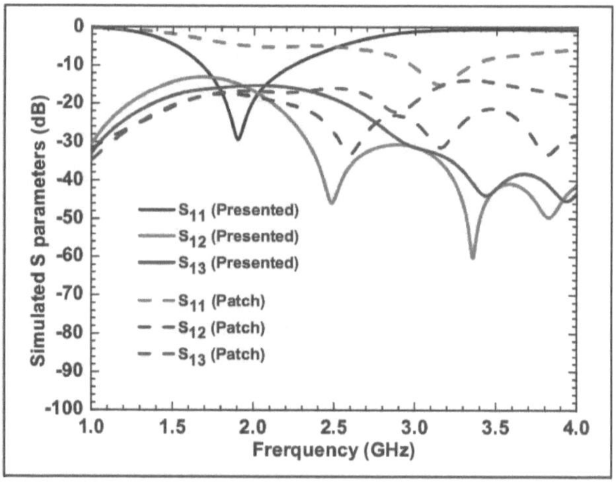

FIGURE 10.19
MIMO S parameters with considered and conventional patch shapes [247]. [From: Leeladhar Malviya, Rajib K. Panigrahi, and M. V. Kartikeyan, "A low profile planar MIMO antenna with polarization diversity for LTE 1800/1900 applications," Microwave and Optical Technology Letters (MOTL), vol. 59, no. 3, pp. 533–538, March 2017. Reproduced courtesy of the John Wiley & Sons, Ltd.]

this case, is 558.0 MHz. Port to port isolations (adjacent and diagonal) are more than 13.0 dB. The presented MIMO results in 1.26 GHz frequency shift towards the lower-frequency side and difference in bandwidth is 198.0 MHz in comparison with conventional patch shaped MIMO. This shows the better utilization of mathematically inspired MIMO for size and return-loss reduction.

Let us compare the mathematically inspired MIMO with the considered partial ground and full ground. Figure 10.20 shows that with the full ground, considered MIMO does not resonate even in 6.0 GHz frequency range, and 46.0 dB isolation is observed here. Mathematically inspired MIMO with the partial ground has much better return-loss, isolation, bandwidth, and diversity parameters.

The return-loss and isolation parameters of mathematically inspired MIMO has already been discussed in previous paragraphs. Figure 10.21 shows measured 2:1 VSWR band of 1.65–2.26 GHz. MIMO resonates at 1.88 GHz, where −36.17 dB return-loss is achieved. The bandwidth, in this case, is 612.0 MHz. Measured isolations at adjacent ports are more than 13.0 dB, and at diagonal ports, more than 18.0 dB. Due to the port coupling and fabrication errors, slight variations are observable in simulated and measured responses.

To obtain the surface current distribution, un-excited ports are terminated by the 50.0 Ω load. Let port 1 is excited and other ports are terminated by 50.0 Ω for any false triggering. Same process is repeated when other ports are excited. Figure 10.22 shows that major current concentrates at feed line

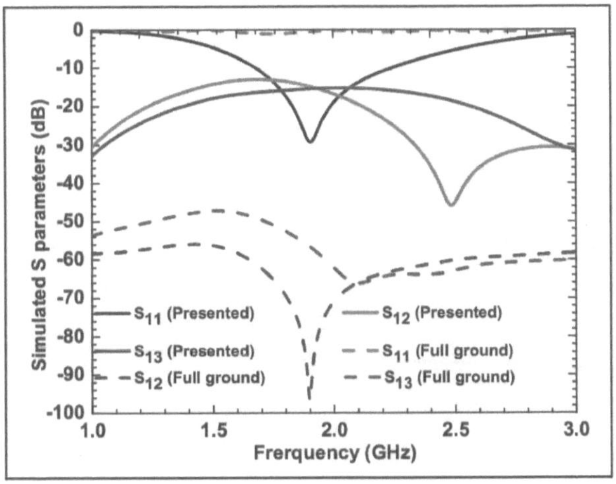

FIGURE 10.20
S parameters with full and partial ground [247]. [From: Leeladhar Malviya, Rajib K. Panigrahi, and M. V. Kartikeyan, "A low profile planar MIMO antenna with polarization diversity for LTE 1800/1900 applications," Microwave and Optical Technology Letters (MOTL), vol. 59, no. 3, pp. 533–538, March 2017. Reproduced courtesy of the John Wiley & Sons, Ltd.]

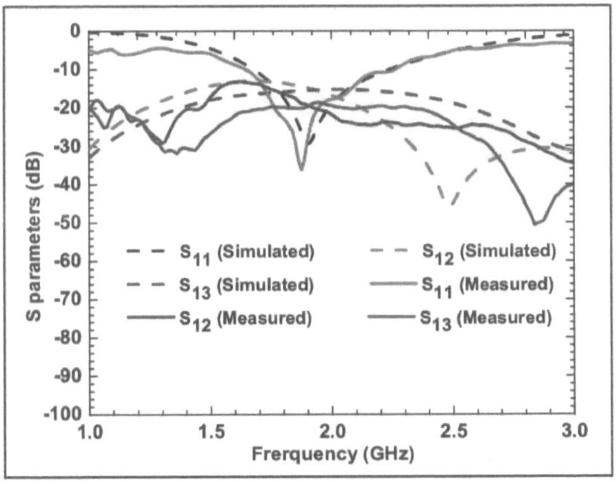

FIGURE 10.21
Simulated and measured S parameters [247]. [From: Leeladhar Malviya, Rajib K. Panigrahi, and M. V. Kartikeyan, "A low profile planar MIMO antenna with polarization diversity for LTE 1800/1900 applications," Microwave and Optical Technology Letters (MOTL), vol. 59, no. 3, pp. 533–538, March 2017. Reproduced courtesy of the John Wiley & Sons, Ltd.]

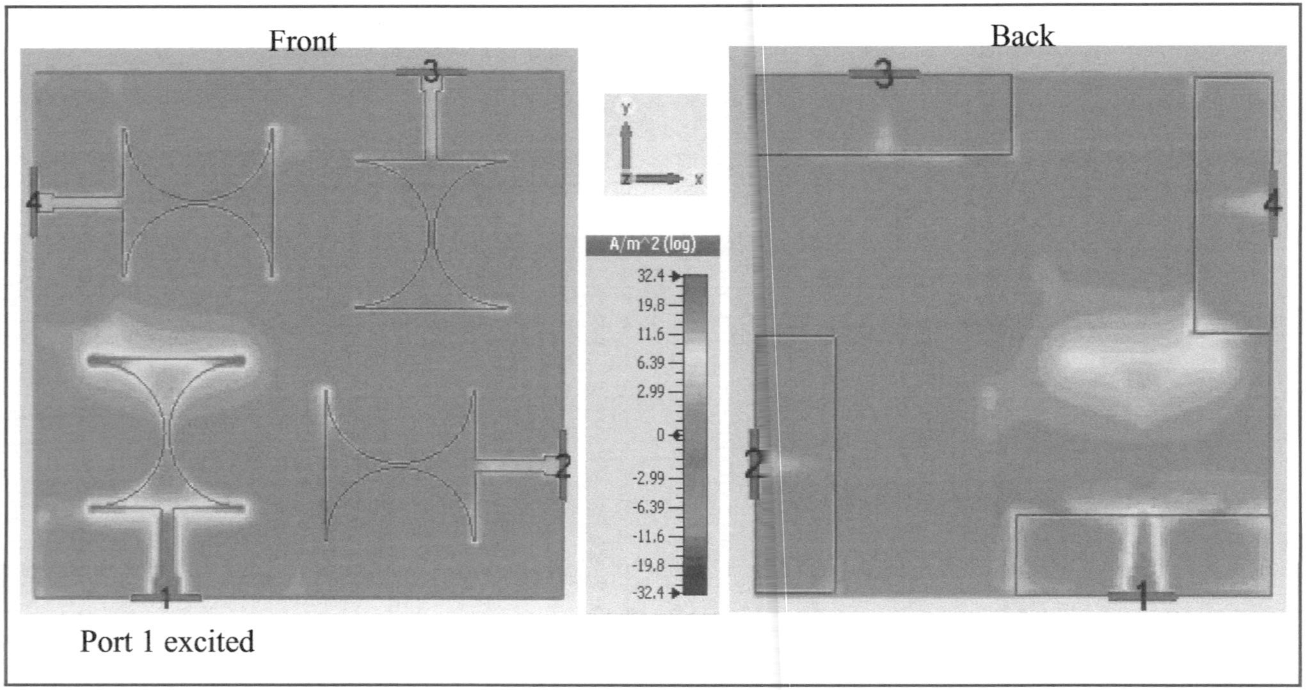

FIGURE 10.22
Current distribution of mathematically inspired MIMO antenna [247]. [From: Leeladhar Malviya, Rajib K. Panigrahi, and M. V. Kartikeyan, "A low profile planar MIMO antenna with polarization diversity for LTE 1800/1900 applications," Microwave and Optical Technology Letters (MOTL), vol. 59, no. 3, pp. 533–538, March 2017. Reproduced courtesy of the John Wiley & Sons, Ltd.]

and edges of port 1 radiating element. Equi-distant ports will link with the equal amount of currents and diagonal elements will link with fewer currents. This can be observed very easily here. Maximum amount of current density at different ports is 32.4 A/m square.

Mathematically inspired patch is a generalized structure, whose effectiveness may also be verified using parametric analysis on its curve radius. A parametric sweep is carried out for the values of r ranging from 9.46 to 12.46 mm. Figure 10.23 shows that for r = 9.46 mm, MIMO radiates at 2.5 GHz frequency and has more than 10.0 dB of isolation among different ports. For r = 11.46 mm (considered radius), MIMO resonates at 1.9 GHz frequency and covers 2:1 VSWR band of 1.68–2.24 GHz, where better return-loss and compactness can be achieved. For r = 12.46 mm, no resonant is observed under 3.0 GHz frequency range. At this radius, the whole shape is divided into two parts.

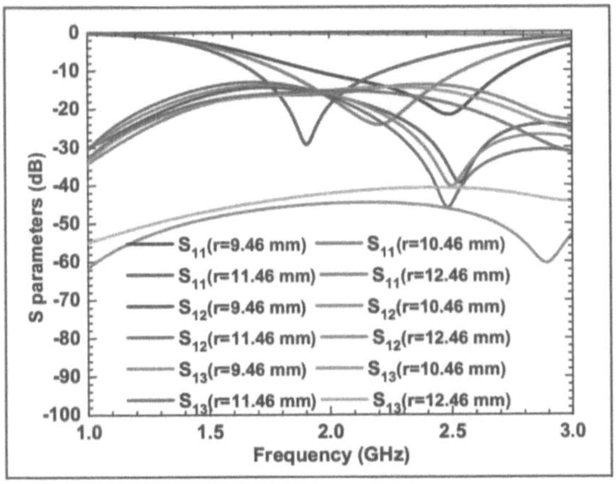

FIGURE 10.23
MIMO S parameters with radius of curvature (r) [247]. [From: Leeladhar Malviya, Rajib K. Panigrahi, and M. V. Kartikeyan, "A low profile planar MIMO antenna with polarization diversity for LTE 1800/1900 applications," Microwave and Optical Technology Letters (MOTL), vol. 59, no. 3, pp. 533–538, March 2017. Reproduced courtesy of the John Wiley & Sons, Ltd.]

Figure 10.24 shows the comparison of equivalent circuit responses with CST-MWS. Return-loss and isolation parameters (S_{11}, S_{12}, and S_{13}) of equivalent circuit and CST-MWS have very minor differences here.

Mathematically inspired MIMO with gain, ECC, TARC, and MEG show the diversity performance for the effectiveness of design. Figure 10.25 shows the comparison of far-field gain and is measured using the substitution method in an anechoic chamber. More than 3.0 dBi simulated gain is observed here in the whole band, whereas, at 1.9 GHz resonant frequency, it is 3.84 dBi.

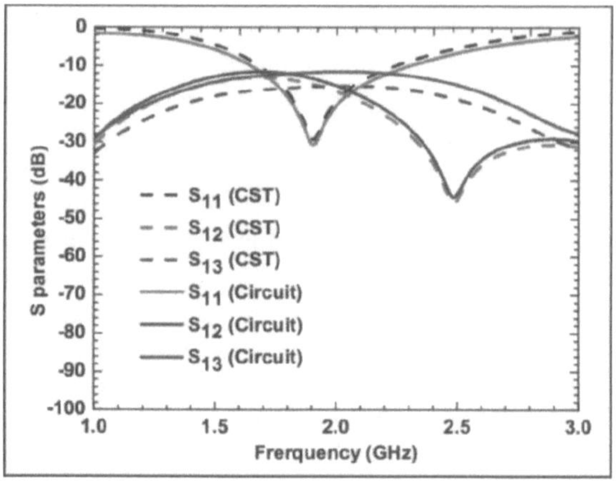

FIGURE 10.24
Equivalent circuit and CST S parameters [247]. [From: Leeladhar Malviya, Rajib K. Panigrahi, and M. V. Kartikeyan, "A low profile planar MIMO antenna with polarization diversity for LTE 1800/1900 applications," Microwave and Optical Technology Letters (MOTL), vol. 59, no. 3, pp. 533–538, March 2017. Reproduced courtesy of the John Wiley & Sons, Ltd.]

Similarly, measured far-field gain in the band is more than 2.8 dBi, and at resonant, it is 3.7 dBi. More than 95.5% radiation efficiency and more than 70% total efficiency are observed in band. At resonant, radiation efficiency is 96.6%.

To obtain the ECC of the mathematically inspired MIMO, let $i = 1$ to 2, $j = 1$ to 2 for any two elements, and $N = 2$ as total 2 antenna elements are selected in each case. Also, substitute $S_{11} = S_{22} = S_{33} = S_{44}$, $S_{12} = S_{21} = S_{34} = S_{43} = S_{14} = S_{41} = S_{23} = S_{32}$, and $S_{13} = S_{31} = S_{24} = S_{42}$ in (4.7). As observable from Fig. 10.26, the simulated and measured ECC lie in the range of 0.1. At resonant (1.9 GHz), simulated ρ_{e12}, ρ_{e13}, and ρ_{e14} are 8.12×10^{-5}, 7.8×10^{-6}, and 8.12×10^{-5}. Similarly, the measured ρ_{e12}, ρ_{e13}, and ρ_{e14} at resonant are 1.35×10^{-7}, 1.9×10^{-6}, and 1.35×10^{-7}. Very low values of ECC in simulated and measured bands are obtained.

Mathematically inspired MIMO diversity performance in terms of TARC is obtained using (4.10). To obtain the TARC, port excitations with different phase angles are considered. Figure 10.27 shows the best case condition for phase relation of 90°, 180° at adjacent/diagonal ports. Other combinations of phase relations lead to the distortions and frequency shifts. For the best phase relation, corresponding TARC curve has resemblance with return-loss parameter (it is not necessary that the shape should be exact) is seen. The TARC bandwidth in each case is better than 500.0 MHz.

Another diversity parameter for the mathematically inspired MIMO is MEG, and is obtained for Isotropic and Gaussian mediums using (4.15). The mean and variance of the Gaussian medium for both vertical and horizontal

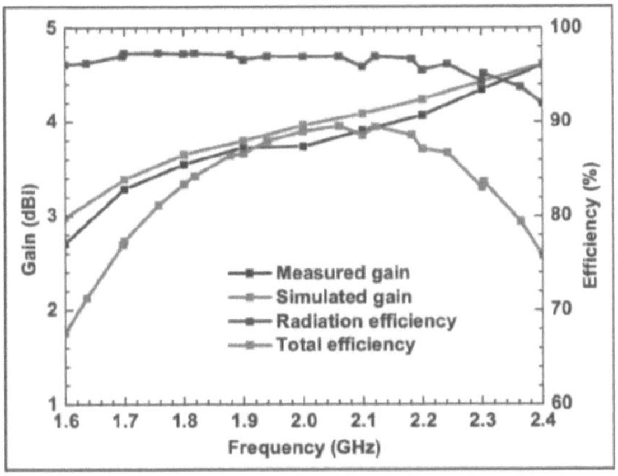

FIGURE 10.25
Gain and efficiency of mathematically inspired MIMO antenna [247]. [From: Leeladhar Malviya, Rajib K. Panigrahi, and M. V. Kartikeyan, "A low profile planar MIMO antenna with polarization diversity for LTE 1800/1900 applications," Microwave and Optical Technology Letters (MOTL), vol. 59, no. 3, pp. 533–538, March 2017. Reproduced courtesy of the John Wiley & Sons, Ltd.]

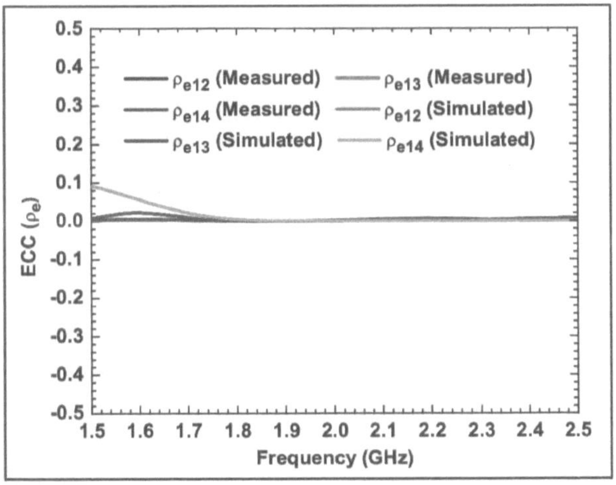

FIGURE 10.26
ECC of mathematically inspired MIMO antenna [247]. [From: Leeladhar Malviya, Rajib K. Panigrahi, and M. V. Kartikeyan, "A low profile planar MIMO antenna with polarization diversity for LTE 1800/1900 applications," Microwave and Optical Technology Letters (MOTL), vol. 59, no. 3, pp. 533–538, March 2017. Reproduced courtesy of the John Wiley & Sons, Ltd.]

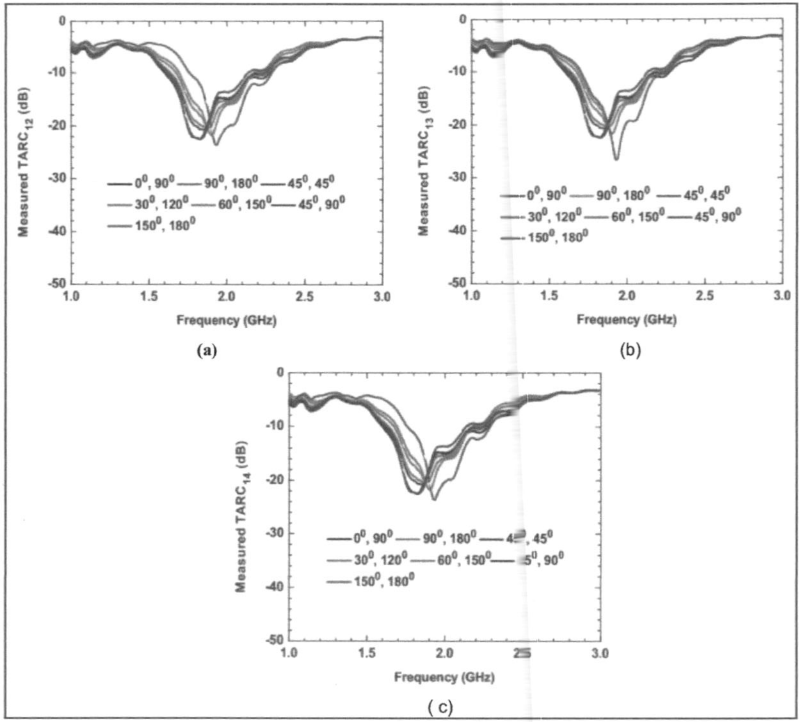

FIGURE 10.27
TARC of the mathematically inspired MIMO antenna [247]. [From: Leeladhar Malviya, Rajib K. Panigrahi, and M. V. Kartikeyan, "A low profile planar MIMO antenna with polarization diversity for LTE 1800/1900 applications," Microwave and Optical Technology Letters (MOTL), vol. 59, no. 3, pp. 533–538, March 2017. Reproduced courtesy of the John Wiley & Sons, Ltd.]

components in this analysis are, mean (μ) = 0 and variance (σ) = 20, for different values of *XPR* for indoor and outdoor activities. Figure 10.28 shows that, for the Isotropic medium and for *XPR* = 0 dB, MEG is constant in frequency range, whereas for *XPR* = 6 dB, large variation is observed. At resonant, −3.0 dB and −4.45 dB MEG values are obtained. For Gaussian medium, for *XPR* = 0 dB and *XPR* = 6 dB, MEG values are <−3.0 dB. At resonant, these are −4.05 dB and −6.96 dB. Due to different MEG values, MIMO fits for the outdoor Isotropic and Gaussian mediums perfectly.

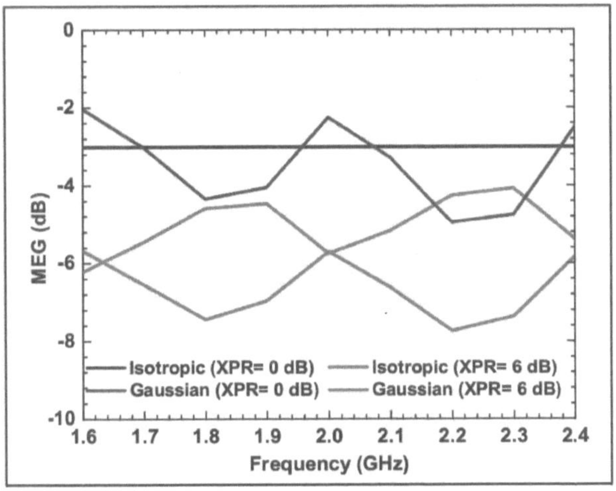

FIGURE 10.28
MEG of mathematically inspired MIMO antenna [247]. [From: Leeladhar Malviya, Rajib K. Panigrahi, and M. V. Kartikeyan, "A low profile planar MIMO antenna with polarization diversity for LTE 1800/1900 applications," Microwave and Optical Technology Letters (MOTL), vol. 59, no. 3, pp. 533–538, March 2017. Reproduced courtesy of the John Wiley & Sons, Ltd.]

Figure 10.29 shows the E-field patterns, where main lobe directions for four radiators are 255°, 190°, 105°, and 170° (for $\phi = 0°$). Also, the beam widths are 182.1°, 79.5°, 182.1°, and 79.5° obtained here. Figure 10.30 shows the H-field radiation patterns, where main lobe directions for four radiators are 170°, 255°, 190°, and 105° (for $\phi = 90°$). The beam widths of each radiator in H-field patterns are 79.5°, 182.1°, 79.5°, and 182.1°. Both the simulated and measured E-field and H-field radiation patterns are broad and able to cover all directions.

Table 10.4 shows the comparison of mathematically inspired MIMO with earlier reported works. The considered references [244] and [245] are compact designs for two elements only. Reference [246] shows good performance at the cost of large size. Therefore, the mathematically inspired MIMO has good diversity behaviour and far-field parameters in terms of dimension, ECC, and gain, in comparison with the earlier reported works.

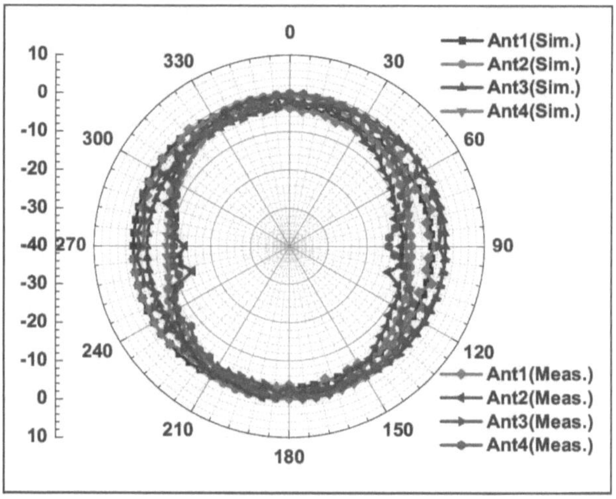

FIGURE 10.29
E-field radiation patterns at resonant [247]. [From: Leeladhar Malviya, Rajib K. Panigrahi, and M. V. Kartikeyan, "A low profile planar MIMO antenna with polarization diversity for LTE 1800/1900 applications," Microwave and Optical Technology Letters (MOTL), vol. 59, no. 3, pp. 533–538, March 2017. Reproduced courtesy of the John Wiley & Sons, Ltd.]

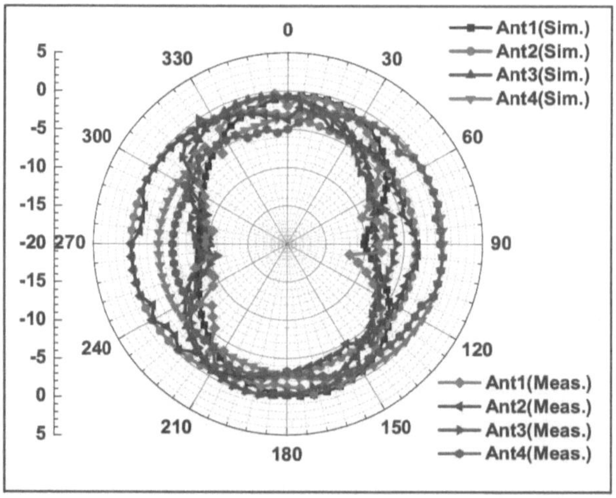

FIGURE 10.30
H-field radiation patterns at resonant [247]. [From: Leeladhar Malviya, Rajib K. Panigrahi, and M. V. Kartikeyan, "A low profile planar MIMO antenna with polarization diversity for LTE 1800/1900 applications," Microwave and Optical Technology Letters (MOTL), vol. 59, no. 3, pp. 533–538, March 2017. Reproduced courtesy of the John Wiley & Sons, Ltd.]

TABLE 10.4
Comparison of mathematically inspired MIMO with earlier reported works [247]. [From: Leeladhar Malviya, Rajib K. Panigrahi, and M. V. Kartikeyan, "A low profile planar MIMO antenna with polarization diversity for LTE 1800/1900 applications," Microwave and Optical Technology Letters (MOTL), vol. 59, no. 3, pp. 533–538, March 2017. Reproduced courtesy of the John Wiley & Sons, Ltd.]

Ref. No.	Frequency/ bands (GHz)	No. of elements	Size (mm^2)	Isolation (dB)	ECC	Gain (dBi)
[243]	1.8/1.9/3.5	2	50 × 50	>24	$<10^{-2}$	5.3 (peak)
[244]	1.61–2.75	2	80 × 60	>15	$<10^{-2}$	—
[245]	1.7–2.9	2	110 × 65	>15	$<10^{-2}$	>1.21
[246]	1.6–2.3	2	125 × 125	>30	—	8.0
In this design	1.68–2.24	4	82.7 × 82.7	>13	$<10^{-2}$	>3.0

10.4 Concluding Remarks

Two MIMO antenna designs with all the possible diversity parameters are demonstrated in this chapter for LTE applications. First design with C-shaped folded-loop MIMO and the partial ground was presented with 2:1 VSWR band of 1.66–2.17 GHz, more than 2.5 dBi gain, and better than 10.0 dB isolation at different ports. The design showed its applicability in both the indoor and outdoor isotropic and Gaussian mediums using MEG behavior [240]. In second design, a mathematically inspired MIMO antenna with polarization diversity has been demonstrated to cover 1.68–2.24 GHz frequency range and more than 13.0 dB isolation in band. More than 3.0 dBi gain is obtained in both the simulated and measured bands, with less than 0.01 ECC. MEG behavior of mathematically inspired MIMO showed its applicability for outdoor isotropic and Gaussian environments [247]. The extension of the work can be done to design other MIMO antennas like proximity coupled rings with curved patch antennas [248].

11

MIMO Antenna Designs for WLAN/WiMAX Applications with 1 × 2 Power Divider Arms

Multiple-input multiple-output (MIMO) antennas with high capacity, high data rate, and compactness of the radiators are the prime objective of the MIMO antenna designers. Regular shapes like square, rectangular, and circular are deformed to improve diversity parameters. Variety of radiation patterns are possible in such cases. Offsetting is one of the examples of improving the bandwidth and isolation in MIMO antenna designs. Two MIMO structures with 1 × 2 power divider arms are described in this chapter. In first design, pentagon-shaped radiator with an inverted L-shaped slot is introduced to resonate at 2.4/3.5 GHz and to achieve omni-directional radiation patterns. In the second design, a square patch with inverted L-shaped slot with the partial ground is introduced to resonate at 2.4/3.5 GHz. All the designs are used for non-line-of-sight (NLOS) applications.

11.1 Introduction

MIMO antenna designs with feed structures have matching capabilities. Power divider arms are used to divide the total power in equal arms for wireless applications and can be used for beamforming also. In this chapter, two different patch structures are selected to design two types of MIMO antenna designs with −10 dB return-loss bandwidth criteria. First design uses 1 × 2 feed arms and pentagon-shaped patch to form 2 × 2 MIMO. This design uses offset effect to utilize the diagonal space by shifting feed from the center position. Second design also uses 1 × 2 feed arms to create 2 × 2 MIMO. Hence, by proper phase relations beamforming can be accomplished. Different radiation patterns in various directions accommodate the specific users for transmission and reception. Every aspect of the design is utilized here to verify and justify the requirement of wireless communication.

11.2 Design and Analysis of Pentagon-Shaped Offset Planar MIMO for Omni-Directional Radiation Patterns

In this section, multi-band MIMO antenna design with pentagon-shaped radiator and offset effect is presented for 2.4/3.5 GHz wireless applications.

11.2.1 Introduction and Related Work

Compactness of radiating elements is the prime requirement of wireless devices. But the closeness of radiating elements in a compact space leads to degradation of design parameters. Mutual coupling reduction techniques improve performance by controlling the amount of current linking or diverting the current in other directions. Hence, better utilization of space with enhanced diversity parameters can be achieved. Combination of different diversity techniques and feed arms have a positive impact on diversity parameters [87, 163]. Variety of radiation patterns are obvious, due to the use of distinctly shaped radiators in multi-element antenna. Asymmetry in radiating elements leads to enhanced link reliability and improved isolation [249, 250]. Slot etching in ground/patch results in multi-band operation and produces an appropriate amount of isolation for LTE/WLAN/WiMAX bands [251]. Metamaterial inspired shapes like SRR/CSRR offers compactness and enhances in-band isolation [201, 252–254].

Different types of power divider arm based antennas are available [255]. In this design, a pentagon-shaped MIMO with inverted L-shaped slot and with T-shaped isolator is designed for omni-directional radiation patterns. Pentagon-shaped MIMO radiator with 1×2 power divider arms resonates at 2.45/3.5 GHz frequencies. The 2:1 VSWR bandwidth is higher than 280.0 MHz in each band. Offset effect is introduced here to utilize diagonal space in the design. Better than 12.4 dB isolation at the ports is achieved in the design.

11.2.2 MIMO Antenna Design and Implementation

CST-MWS optimized MIMO antenna with pentagon-shaped radiators, inverted L-shaped slots and T-shaped isolator is presented with 50.0 Ω ports. In diagonal space, T-shaped isolating arms are designed for space utilization to achieve low mutual coupling. CST-MWT designed MIMO uses FR-4 dielectric substrate of 1.524 mm thickness, permittivity of 4.4, and loss tangent of 0.025, and has 65.3×65.3 mm^2 area.

The conventional patch shows slightly less performance in comparison with a pentagon-shaped patch. Pentagon-shaped MIMO with 1×2 power divider arms achieves omni-directional performance. The T-shaped isolator provides

TABLE 11.1
CST-MWS optimized parameters of pentagon-shaped offset MIMO (mm) [256] [From: Leeladhar Malviya, M. V. Kartikeyan, and Rajib K. Panigrahi, "Offset planar MIMO antenna for omnidirectional radiation patterns," International Journal of RF and Microwave Computer Aided Engineering, pp. 1–9, October 2018. Reproduced courtesy of the John Wiley & Sons, Ltd.]

Parameter	a	b	c	d	e	f	g	h	i	j
Value	65.3	65.3	27.53	11.17	8.97	15.05	0.8	10.35	6.43	2.96
Parameter	k	l	m = n	o	p	q	r	s	t	u
Value	16.36	4.68	17.18	6.87	0.57	0.7	39.18	0.57	12.25	8.35

control on the mutual coupling. The frequency of operation = 2.4/3.5 GHz, VSWR = 2:1, isolation >10 dB, gain (dBi) >5, efficiency (%) >90, ECC <0.1, and WLAN/WiMAX applications are the set goals of the MIMO antenna design shown here. Table 11.1 shows the optimized design parameters of the presented MIMO. Figures 11.1 and 11.2 show the schematic and fabricated MIMO antenna views.

Tuning property by an equivalent circuit is presented using ADS software for 50.0 Ω ports. CST-MWS has the facility of extracting the values of these parameters. Due to the basic definitions of the inductor and capacitor, metallic, non-metallic, and overlapping areas are represented by equivalent inductor and capacitor components of the designed antenna. Parallel combination of L1–C1 represents the first resonant peak, and the second resonant peak is represented by the parallel combination of L2–C2. Isolation parameters of the presented MIMO (S_{12} or S_{21}) are represented by the parallel combination of L3–C3. Figure 11.3 shows the circuit parameters as: L1 = 0.816 nH, C1 = 5.24 pf, L2 = 5.98 nH, C2 = 0.6 pf, L3 = 3.98 nH, and C3 = 0.46 pf. Each of the port is represented by the 50.0 Ω resistances.

11.2.3 Simulation-Measurement Results and Discussion

Performance parameters of the pentagon-shaped MIMO are evaluated and validated with the help of VNA-HP8720B and in an anechoic chamber. From CST simulation and VNA-HP8720B, it is observed that port return-losses S_{11} = S_{22}, and isolation S_{12} = S_{21}. Let us consider only S_{11} and S_{12} scattering parameters for the analysis of other results of the design. With full ground, the presented pentagon-shaped offset MIMO does not resonant in 4.5 GHz frequency range. The presented pentagon-shaped offset MIMO resonates in 2:1 VSWR bands of 2.24–2.64 GHz, and 3.41–3.69 GHz frequencies. Simulated percentage bandwidths in lower and higher bands are 16.39% and 7.88% respectively. Isolation in these bands is better than 12.4 dB. Measured 2:1 VSWR bandwidth in lower and higher bands are 20.34% (2.12–2.60 GHz) and 8.15% (3.41–3.70 GHz). Measured in-band isolation in both the bands is better than 14.0 dB. All the simulated and measured results are in good agreement and are observable in Fig. 11.4. Differences in bandwidths and isolations are due to fabrication errors and port/cable coupling losses.

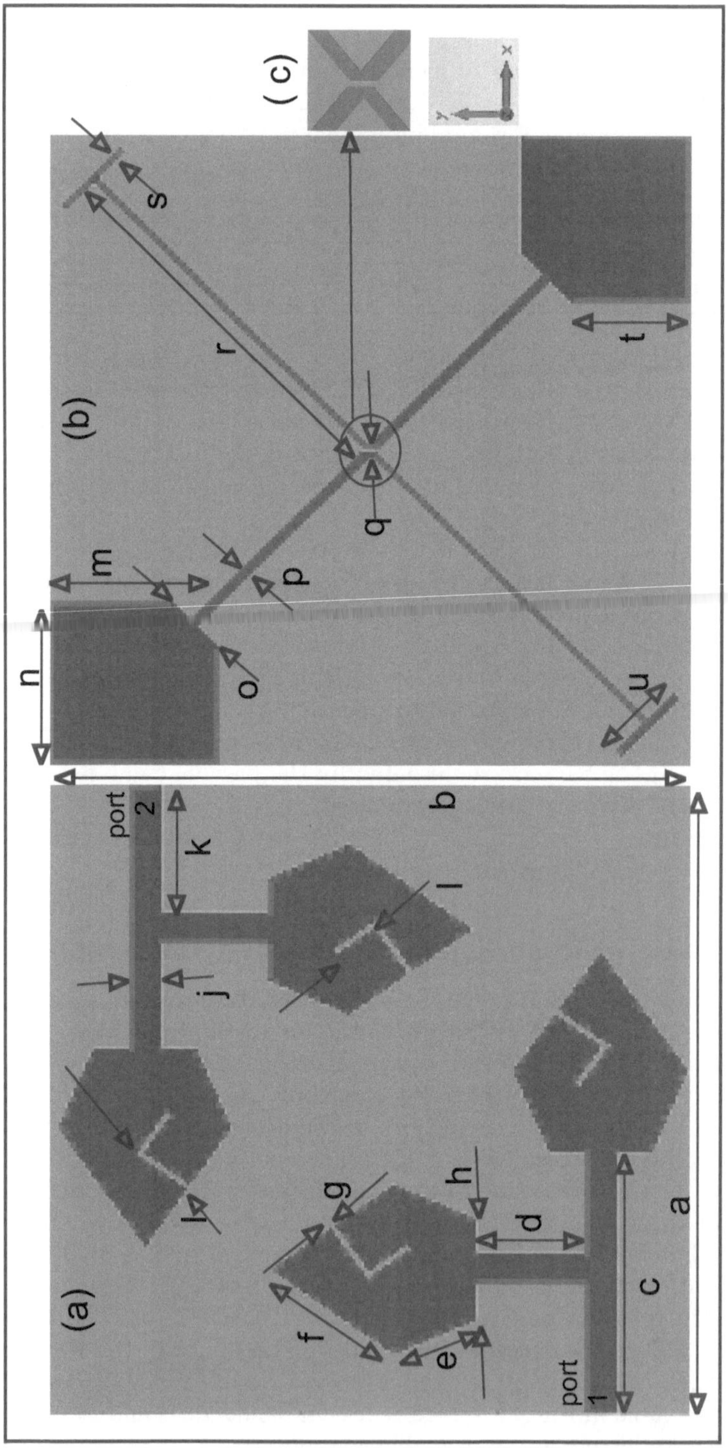

FIGURE 11.1
Schematic views of pentagon-shaped offset MIMO [256]. [From: Leeladhar Malviya, M. V. Kartikeyan, and Rajib K. Panigrahi, "Offset planar MIMO antenna for omnidirectional radiation patterns," International Journal of RF and Microwave Computer Aided Engineering, pp. 1–9, October 2018. Reproduced courtesy of the John Wiley & Sons, Ltd.]

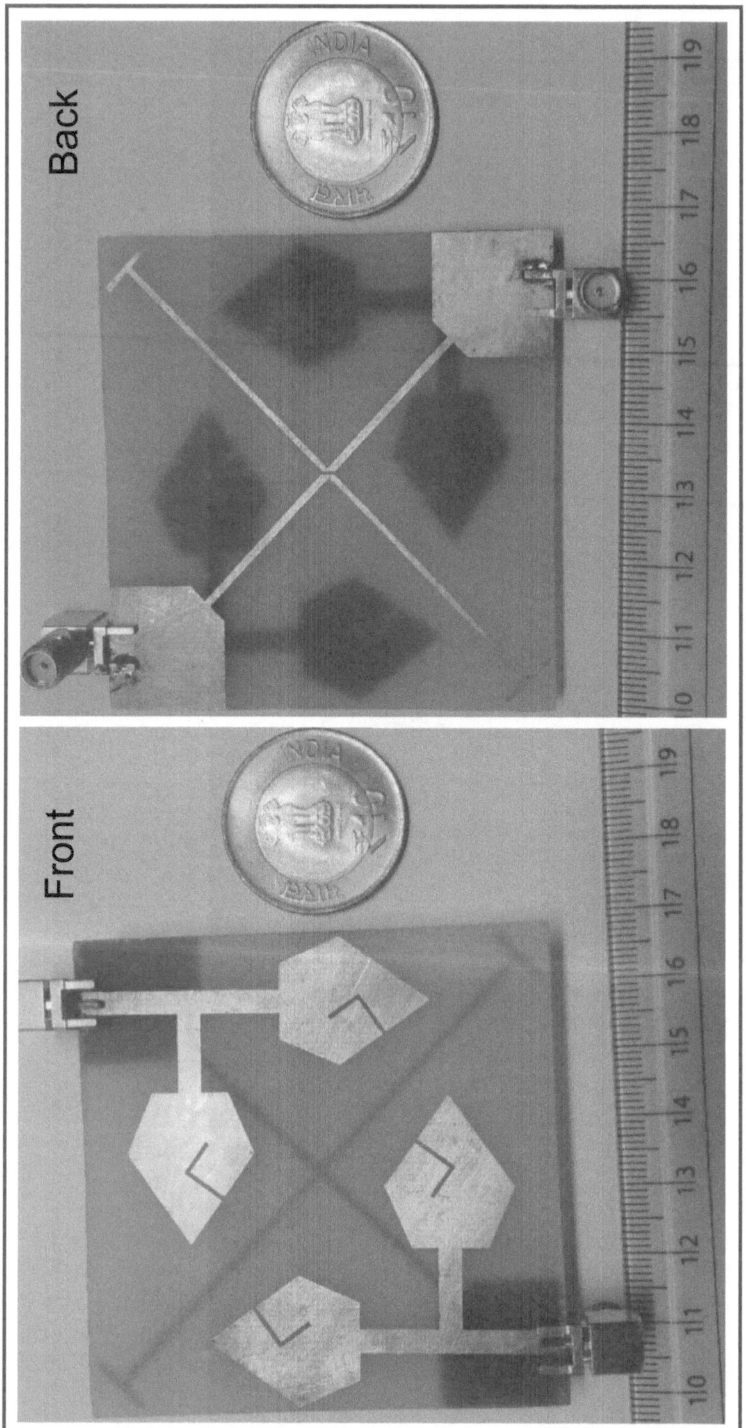

FIGURE 11.2
Fabricated views of pentagon-shaped offset MIMO [256]. [From: Leeladhar Malviya, M. V. Kartikeyan, and Rajib K. Panigrahi, "Offset planar MIMO antenna for omnidirectional radiation patterns," International Journal of RF and Microwave Computer Aided Engineering, pp. 1–9, October 2018. Reproduced courtesy of the John Wiley & Sons, Ltd.]

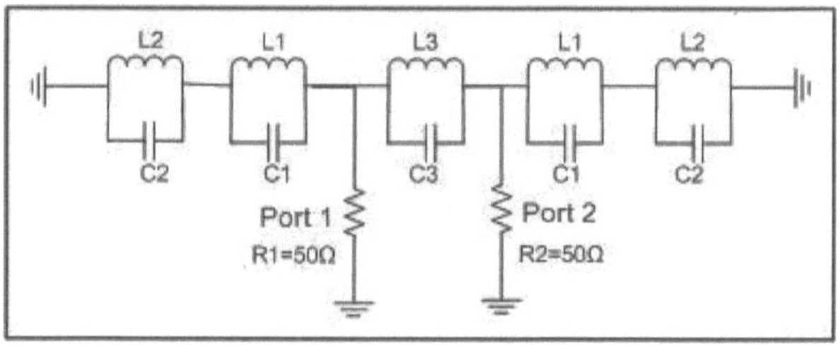

FIGURE 11.3
Equivalent circuit of pentagon-shaped offset MIMO [256]. [From: Leeladhar Malviya, M. V. Kartikeyan, and Rajib K. Panigrahi, "Offset planar MIMO antenna for omnidirectional radiation patterns," International Journal of RF and Microwave Computer Aided Engineering, pp. 1–9, October 2018. Reproduced courtesy of the John Wiley & Sons, Ltd.]

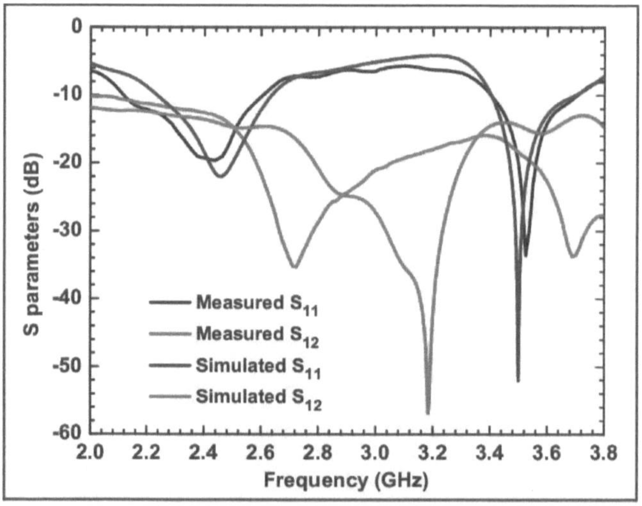

FIGURE 11.4
S parameters of pentagon-shaped offset MIMO [256]. [From: Leeladhar Malviya, M. V. Kartikeyan, and Rajib K. Panigrahi, "Offset planar MIMO antenna for omnidirectional radiation patterns," International Journal of RF and Microwave Computer Aided Engineering, pp. 1–9, October 2018. Reproduced courtesy of the John Wiley & Sons, Ltd.]

Figure 11.5 shows the effect of surface current density on the presented MIMO. Isolation level is less than 9.0 dB, in the absence of T-shaped isolator. Presence of T-shaped isolator enhances isolation up to 12.4 dB in lower band. Figure 11.5 (a) shows concentrated current at 2.45 dB on divider arms when port 1 is excited and port 2 is terminated by 50.0 Ω. Direct current coupling of port 1 is restricted due to the presence of T-shaped arm with port 2. Figure 11.5 (b) shows concentrated current at 3.5 GHz resonant frequency, on patch only. Hence, T-shaped arm has great influence on the lower arm. Same action may be observed when port 2 is excited. The longer current path is established due to the T-shaped arm, and a lower-frequency band is achieved.

Variety of isolation approaches have distinct effects on isolation. Parasitic element may be used in design, but impedance match may become complex, due to inductive effect. T-shaped isolator uses diagonal space without further space requirements. Similarly, use of SRR leads to the bandwidth shrinking, due to inductive-capacitive effect. Figure 11.6 shows the comparison of S parameters with and without T-shaped arm. Better response curves are observed in presence of T-shaped arm.

Parametric evaluation on certain critical parameters is carried out to show the effectiveness of the presented design. Parametric variation on slot width g is done in the range of 0.2–1.0 mm. Favorable S_{11} and S_{12} responses are observed for slot width equals 0.8 mm, at 2.45/3.5 GHz resonant frequencies. Figure 11.7 shows the variation in S parameters with slot width g. Very low (−54 dB) return-loss is observable for the considered slot width. There are no significant changes in isolation parameter with different values of g.

In the same manner, slot length l is varied in the range of 2.68–6.68 mm. For considered length l = 4.68 mm, favorable S parameters are observed in Fig. 11.8. Variation in isolation is observed here for different values of l.

Similarly, the length r of the diagonal arm is varied to observe changes in S parameters in Fig. 11.9. For the considered length r = 39.92 mm, best results are evidenced. Other values of r result in minor variations in both the return-loss and isolation parameters.

Diversity parameters for the pentagon-shaped MIMO are discussed in terms of gain, ECC, and TARC of the fabricated prototype. Figure 11.10 shows the measured and simulated gain comparison along with radiation and total efficiencies. Far-field gain is measured in an anechoic chamber with the substitution method. CST simulated gain and measured gain are greater than 2.6 dBi in lower and higher bands, and at 2.45 GHz and 3.5 GHz frequencies, gains are 3.2 dBi and 4.77 dBi, respectively. At 2.43 GHz and 3.52 GHz, resonant frequencies the measured gains are 3.2 dBi and 4.52 dBi, respectively. Similarly, the radiation efficiency in each band is more than 71.5%. Total efficiency is also more than 50% in each band.

ECC of the presented MIMO is obtained using (4.7). Figure 11.11 shows the comparison of simulated and measured ECCs. In both the simulated bands

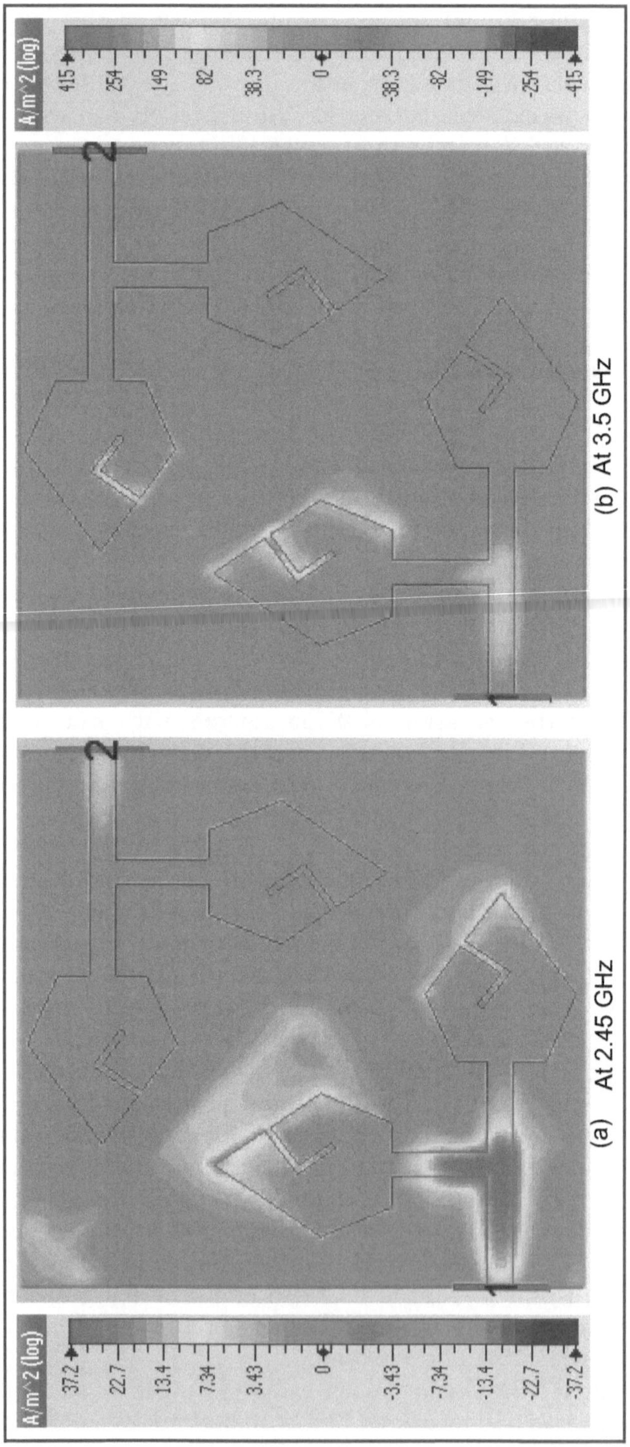

FIGURE 11.5
Current distribution of pentagon-shaped offset MIMO [256]. [From: Leeladhar Malviya, M. V. Kartikeyan, and Rajib K. Panigrahi, "Offset planar MIMO antenna for omnidirectional radiation patterns," International Journal of RF and Microwave Computer Aided Engineering, pp. 1–9, October 2018. Reproduced courtesy of the John Wiley & Sons, Ltd.]

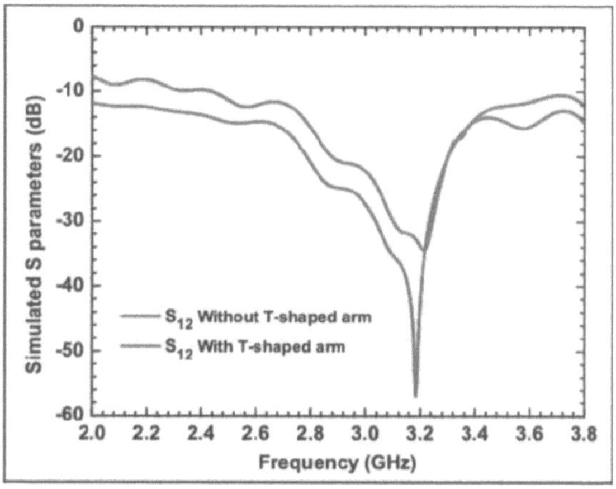

FIGURE 11.6
Isolation with and without T-shaped arm [256]. [From: Leeladhar Malviya, M. V. Kartikeyan, and Rajib K. Panigrahi, "Offset planar MIMO antenna for omnidirectional radiation patterns," International Journal of RF and Microwave Computer Aided Engineering, pp. 1–9, October 2018. Reproduced courtesy of the John Wiley & Sons, Ltd.]

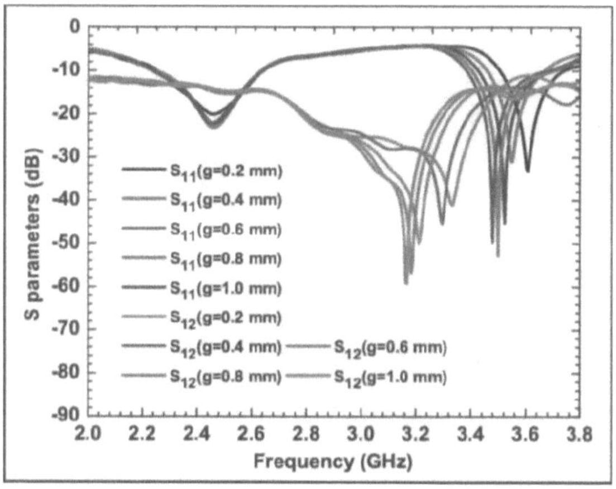

FIGURE 11.7
S parameter variation with slot width g [256]. [From: Leeladhar Malviya, M. V. Kartikeyan, and Rajib K. Panigrahi, "Offset planar MIMO antenna for omnidirectional radiation patterns," International Journal of RF and Microwave Computer Aided Engineering, pp. 1–9, October 2018. Reproduced courtesy of the John Wiley & Sons, Ltd.]

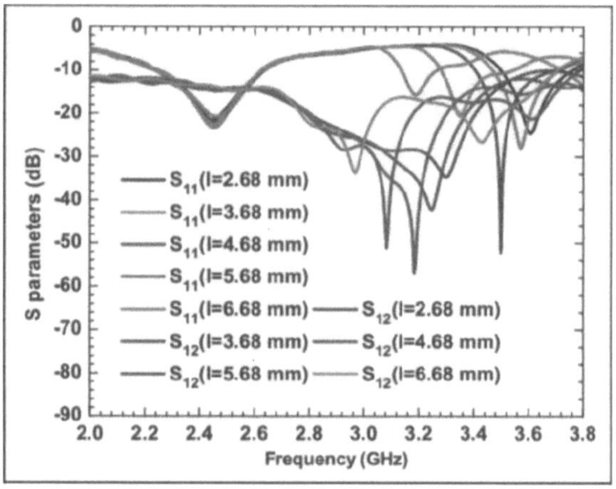

FIGURE 11.8
S parameter variation with slot length l [256]. [From: Leeladhar Malviya, M. V. Kartikeyan, and Rajib K. Panigrahi, "Offset planar MIMO antenna for omnidirectional radiation patterns," International Journal of RF and Microwave Computer Aided Engineering, pp. 1–9, October 2018. Reproduced courtesy of the John Wiley & Sons, Ltd.]

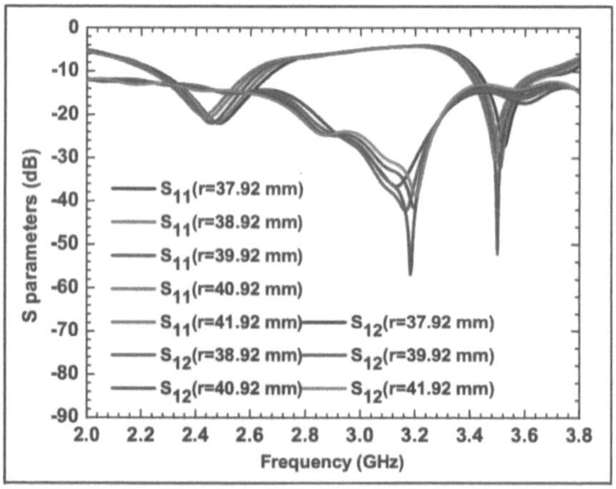

FIGURE 11.9
S parameter variation with length r [256]. [From: Leeladhar Malviya, M. V. Kartikeyan, and Rajib K. Panigrahi, "Offset planar MIMO antenna for omnidirectional radiation patterns," International Journal of RF and Microwave Computer Aided Engineering, pp. 1–9, October 2018. Reproduced courtesy of the John Wiley & Sons, Ltd.]

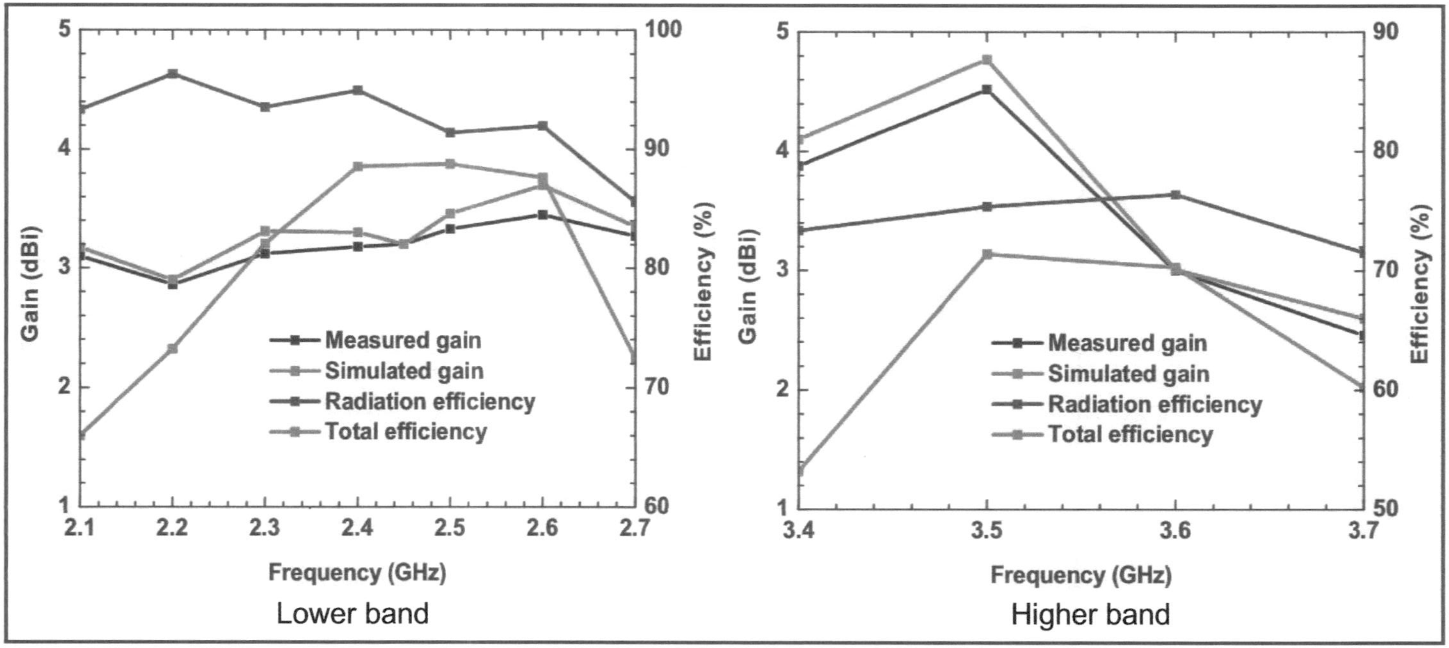

FIGURE 11.10
Gain and efficiency of pentagon-shaped offset MIMO antenna [256]. [From: Leeladhar Malviya, M. V. Kartikeyan, and Rajib K. Panigrahi, "Offset planar MIMO antenna for omnidirectional radiation patterns," International Journal of RF and Microwave Computer Aided Engineering, pp. 1–9, October 2018. Reproduced courtesy of the John Wiley & Sons, Ltd.]

ECC is less than 0.01, while measured values are slightly higher than simulated values. A very low value of ECC in CST-MWS and measured bands is an advantage of the presented design.

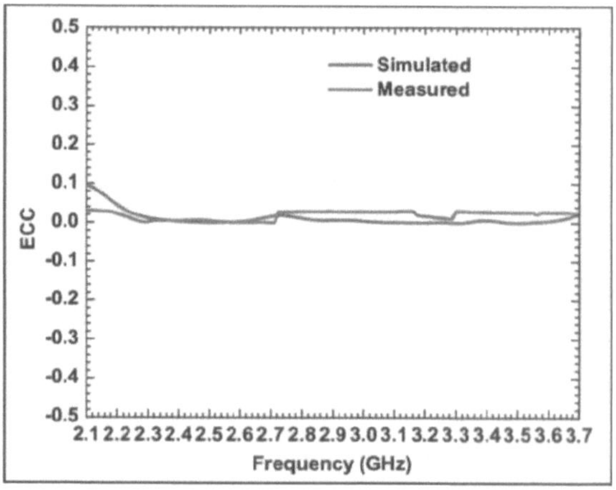

FIGURE 11.11
ECC of pentagon-shaped offset MIMO antenna [256]. [From: Leeladhar Malviya, M. V. Kartikeyan, and Rajib K. Panigrahi, "Offset planar MIMO antenna for omnidirectional radiation patterns," International Journal of RF and Microwave Computer Aided Engineering, pp. 1–9, October 2018. Reproduced courtesy of the John Wiley & Sons, Ltd.]

Diversity performance in terms of TARC is carried out for the presented design in presence of random signals and their phase excitations at ports and is obtained using (4.10). Figure 11.12 shows the comparison of different pair of phase excitations at ports. Best phase excitation combination with $0°$, $90°$ resemblance the shape of return-loss parameter at port 1. For other phase combinations, variations are observed. The active TARC bandwidth with respect to 2:1 VSWR is more than 280 MHz in each band.

Similarly, diversity parameter MEG for indoor and outdoor activities with isotropic and Gaussian mediums are obtained using (4.15). Let horizontal and vertical components of the Gaussian signals have mean $(\mu) = 0$ and variance $(\sigma) = 20$, respectively. Figure 11.13 compares MEG responses for outdoor environment ($XPR = 0$ dB) and indoor environment ($XPR = 6$ dB). Pentagon-shaped offset MIMO has MEG ≤ -3 dB in the whole 2:1 VSWR band for Gaussian medium, and also ≤ -3 dB for Isotropic medium. Hence, due to MEG for these mediums, pentagon-shaped MIMO is applicable for both the indoor and outdoor activities.

Similarly, the CST-MWS responses and equivalent circuit responses of presented pentagon-shaped offset MIMO are compared in Fig. 11.14. It is observed that the CST and ADS responses have very small variations, and are in good agreement.

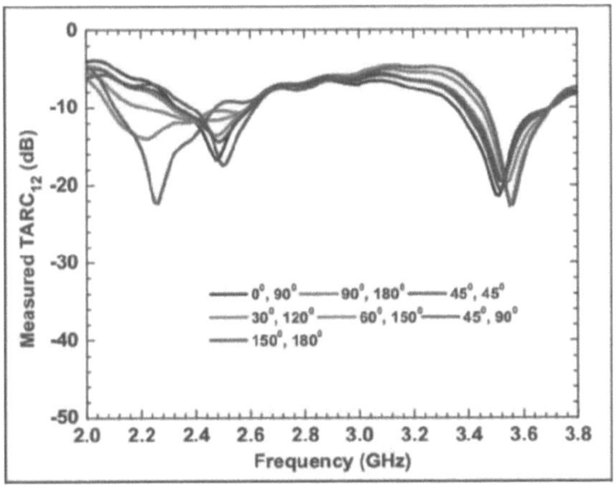

FIGURE 11.12
TARC of pentagon-shaped offset MIMO antenna [256]. [From: Leeladhar Malviya, M. V. Kartikeyan, and Rajib K. Panigrahi, "Offset planar MIMO antenna for omnidirectional radiation patterns," International Journal of RF and Microwave Computer Aided Engineering, pp. 1–9, October 2018. Reproduced courtesy of the John Wiley & Sons, Ltd.]

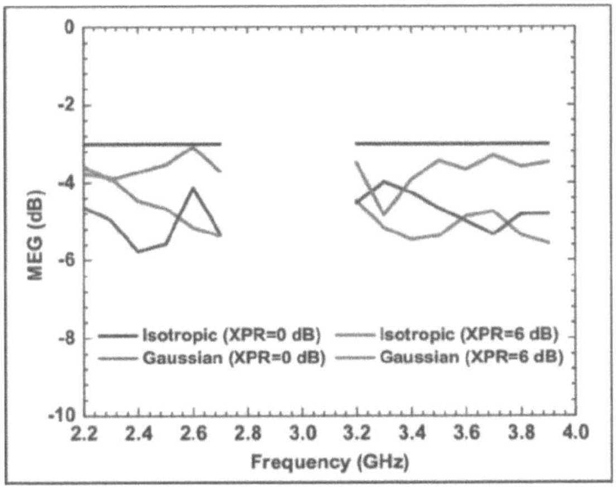

FIGURE 11.13
MEG of pentagon-shaped offset MIMO antenna [256]. [From: Leeladhar Malviya, M. V. Kartikeyan, and Rajib K. Panigrahi, "Offset planar MIMO antenna for omnidirectional radiation patterns," International Journal of RF and Microwave Computer Aided Engineering, pp. 1–9, October 2018. Reproduced courtesy of the John Wiley & Sons, Ltd.]

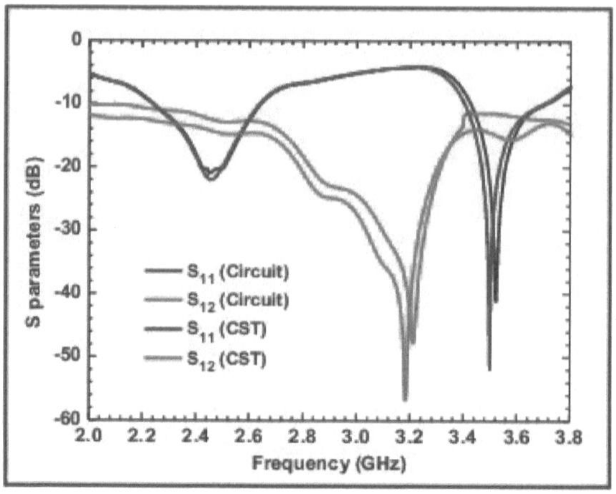

FIGURE 11.14
S parameters of equivalent circuit and CST of MIMO antenna [256]. [From: Leeladhar Malviya, M. V. Kartikeyan, and Rajib K. Panigrahi, "Offset planar MIMO antenna for omnidirectional radiation patterns," International Journal of RF and Microwave Computer Aided Engineering, pp. 1–9, October 2018. Reproduced courtesy of the John Wiley & Sons, Ltd.]

Far-field radiation patterns are measured in an anechoic chamber in the presence of standard horn antennas and power meter for the validation of the pentagon-shaped offset MIMO antenna. Figures 11.15, 11.16, 11.17, and 11.18 show the E-field and H-field radiation patterns and are wide to cover the maximum area. Main lobe directions of E-field for two ports are 205° and 155°, and beam widths at each ports is 277.9°. Similarly, the main lobe directions for H-field at two ports are 205° and 155°. The beam widths at each ports is 277.9°.

Table 11.2 compares the presented design and earlier reported designs. It is observed that slightly better performance is achieved with the presented design in comparison with the earlier reported works in terms of size, isolation, ECC, and gain.

11.3 Design and Analysis of Multi-Band MIMO with Very Compact Radiating Element

The second design in this chapter is also the multi-band compact MIMO antenna for 2.4/3.5 GHz wireless applications, and it is discussed in this section.

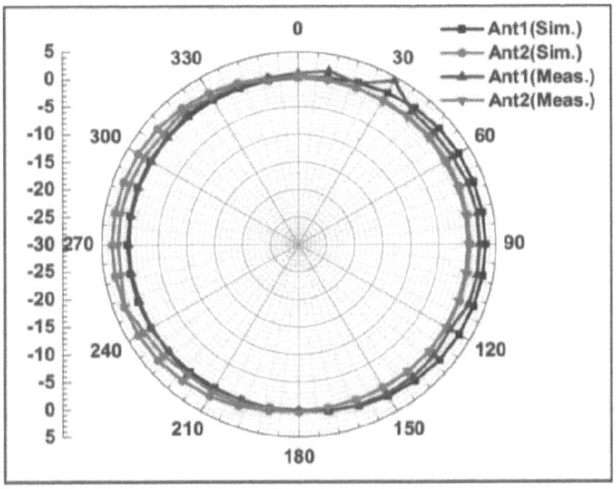

FIGURE 11.15
E-field radiation patterns of presented MIMO antenna at 2.43 GHz [256]. [From: Leeladhar Malviya, M. V. Kartikeyan, and Rajib K. Panigrahi, "Offset planar MIMO antenna for omnidirectional radiation patterns," International Journal of RF and Microwave Computer Aided Engineering, pp. 1–9, October 2018. Reproduced courtesy of the John Wiley & Sons, Ltd.]

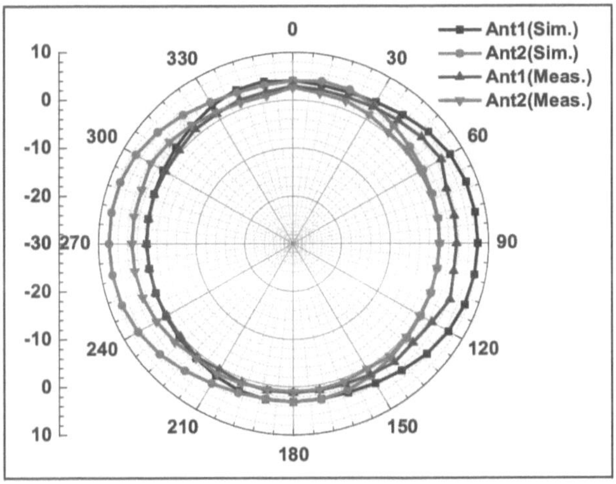

FIGURE 11.16
E-field radiation patterns of presented MIMO antenna at 3.52 GHz [256]. [From: Leeladhar Malviya, M. V. Kartikeyan, and Rajib K. Panigrahi, "Offset planar MIMO antenna for omnidirectional radiation patterns," International Journal of RF and Microwave Computer Aided Engineering, pp. 1–9, October 2018. Reproduced courtesy of the John Wiley & Sons, Ltd.]

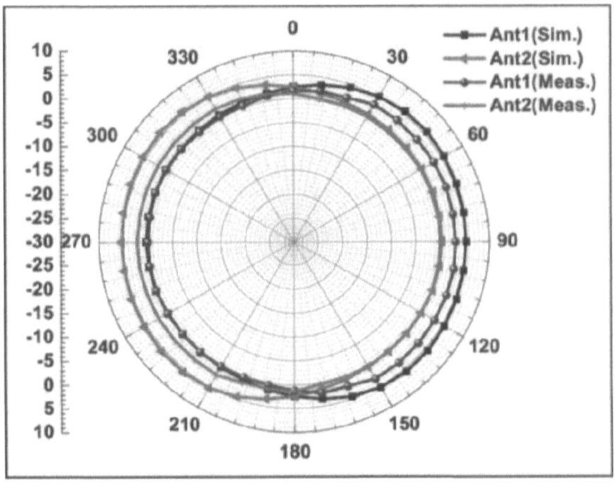

FIGURE 11.17
H-field radiation patterns of presented MIMO antenna at 2.43 GHz [256]. [From: Leeladhar Malviya, M. V. Kartikeyan, and Rajib K. Panigrahi, "Offset planar MIMO antenna for omnidirectional radiation patterns," International Journal of RF and Microwave Computer Aided Engineering, pp. 1–9, October 2018. Reproduced courtesy of the John Wiley & Sons, Ltd.]

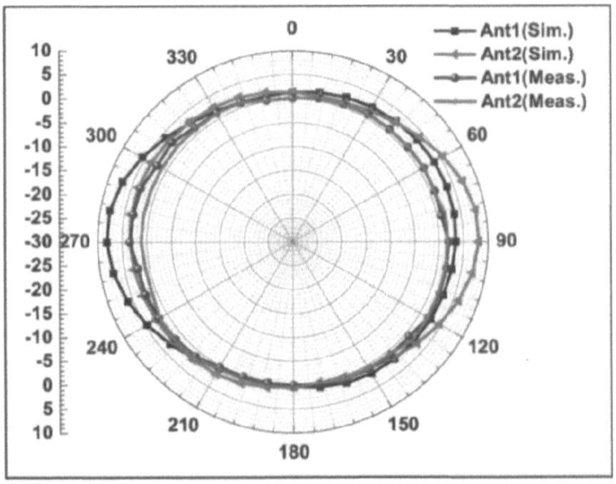

FIGURE 11.18
H-field radiation patterns of presented MIMO antenna at 3.52 GHz [256]. [From: Leeladhar Malviya, M. V. Kartikeyan, and Rajib K. Panigrahi, "Offset planar MIMO antenna for omnidirectional radiation patterns," International Journal of RF and Microwave Computer Aided Engineering, pp. 1–9, October 2018. Reproduced courtesy of the John Wiley & Sons, Ltd.]

TABLE 11.2
Comparison of pentagon-shaped offset MIMO with earlier reported works [256]. [From: Leeladhar Malviya, M. V. Kartikeyan, and Rajib K. Panigrahi, "Offset planar MIMO antenna for omnidirectional radiation patterns," International Journal of RF and Microwave Computer Aided Engineering, pp. 1–9, October 2018. Reproduced courtesy of the John Wiley & Sons, Ltd.]

Ref. No.	Frequency/ bands (GHz)	No. of elements	Size (mm^2)	Isolation (dB)	ECC	Gain (dBi)
[84]	2.65	3	120 × 90	>20	<0.4	2.8
[87]	2.4	2	50 × 50	—	—	—
[94]	2.45	4	100 × 55	>10	<0.01	2.15
[120]	2.4	2	100 × 100	>40	—	3.4
[163]	2.4	2	60 × 90	>29	<0.2	—
[199]	2.4	4	100 × 60	>11	<0.3	>3.12
[254]	2.45	2	100 × 150	>15	$<10^{-3}$	>−0.7
In this design	2.24–2.64 3.41–3.69	4	65.3 × 65.3	>12.4	$<10^{-2}$	>2.6

11.3.1 Introduction and Related Work

MIMO antennas in LTE, WLAN, WiMAX, UWB, and other wireless applications have very compact antenna requirements. Also, the reliability, high data rate, and high capacity are the requirements of these antenna elements. MIMO with compact element results in overall size reduction, but at the cost of extension in mutual coupling. MIMO with enhanced isolation is one of the challenges for antenna designers. As the slot is used for multi-banding and isolation enhancement [80, 94]. EBG and other slow-wave structures are also required for mutual coupling reduction [200, 201]. Multi-user MIMO, and Massive MIMO are the extensions of MIMO antennas [242]. Variety of MIMO designs are used for user-friendly needs [149, 245]. Type of substrate and its loss tangent also decides the overall losses in MIMO [246, 254]. Similarly, offset is also used to maintain the required separation between ports for isolation enhancement [255, 256].

The second design in this chapter is a compact multi-band MIMO for multi-standard applications. This design uses the concept of 1 × 2 feed arms for wireless applications. The very compact size of the radiating element is used to reduce the overall size of implementation here. The presented design is applicable for 2.4/3.5 GHz frequency bands, and has more than 12.5 dB isolation. A unique ground structure is also presented to control adverse effects of mutual coupling.

11.3.2 MIMO Antenna Design and Implementation

MIMO with compact radiators is designed on FR-4 dielectric substrate of thickness 1.524 mm. The frequency of operation = 2.4/3.5 GHz, VSWR = 2:1, isolation >10 dB, gain (dBi) >5, efficiency (%) >90, ECC <0.1, and

WLAN/WiMAX application are the set goals of the MIMO antenna design shown here. Compact radiators and 1 × 2 feed arms and with special ground structure have optimized design parameters given in Table 11.3. Figures 11.19 and 11.20 show the schematic and fabricated MIMO antenna views. Presented MIMO antenna uses the partial ground to add the compactness in the overall structure. Full ground requires a very big size for the same resonant frequencies. Hence, partial ground with ground stub controls the adverse effects of mutual coupling. Without the ground stub, isolation is less than 10 dB.

TABLE 11.3
CST optimized values of compact MIMO antenna (mm) [257]. [From: Leeladhar Malviya, M. V. Kartikeyan, and Rajib K. Panigrahi, "Multi-standard, multi-band planar MIMO antenna with diversity effects for wireless applications," International Journal of RF and Microwave Computer Aided Engineering, pp. 1–8, September 2018. Reproduced courtesy of the John Wiley & Sons, Ltd.]

Parameter	a	b	c	d	e	f	g	h
Value	45.1	90.2	6.7	6.7	30.3	3.0	0.3	5.2
Parameter	i	j	k	l	m	n	o	p
Value	5.3	7.0	4.3	13.2	1.9	51.9	2.4	4.2

Equivalent circuit analysis of the presented MIMO antenna is carried out using ADS software for 50.0 Ω ports. CST-MWS based MIMO design with all the metallic and non-metallic areas are considered to find equivalent circuit parameters. The parallel combination of L1–C1 represents the first frequency band, and the parallel combination of L2–C2 is used for the second resonant band. Isolation parameters S_{12} or S_{21} is represented by the parallel combination of L3–C3. Figure 11.21 shows the circuit parameters as: L1 = 0.82 nH, C1 = 5.24 pf, L2 = 5.98 nH, C2 = 0.6 pf, L3 = 3.98 nH, and C3 = 0.46 pf. Each of the port is represented by the 50.0 Ω resistance.

11.3.3 Simulation-Measurement Results and Discussion

Compact 2×2 MIMO antenna design is fabricated and checked for the validity of return-loss, isolation, and far-field results using VNA and in an anechoic chamber. Symmetry in elements and ports leads to equal return-loss and isolation parameters. Hence, for simplification of discussions, only S_{11} and S_{12} parameters are considered throughout the chapter for this design. Simulated 2:1 VSWR bands of 2.23–2.64 GHz (lower band) and 3.26–3.70 GHz (higher band) are achieved in CST-MWS design. Similarly, measured 2:1 VSWR bands of 2.23–2.63 GHz (lower band) and 3.26–3.69 GHz (higher band) are validated using VNA. Percentage bandwidths in these bands are 16.46% and 12.37%, respectively. Compact MIMO antenna resonates at 2.38/3.5 GHz frequencies.

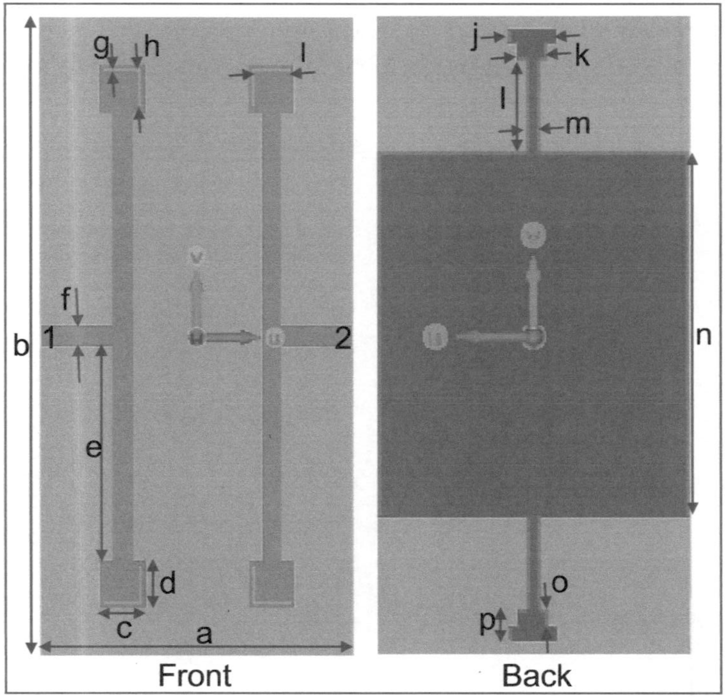

FIGURE 11.19
Schematic views of 2 × 2 MIMO antenna [257]. [From: Leeladhar Malviya, M. V. Kartikeyan, and Rajib K. Panigrahi, "Multi-standard, multi-band planar MIMO antenna with diversity effects for wireless applications," International Journal of RF and Microwave Computer Aided Engineering, pp. 1–8, September 2018. Reproduced courtesy of the John Wiley & Sons, Ltd.]

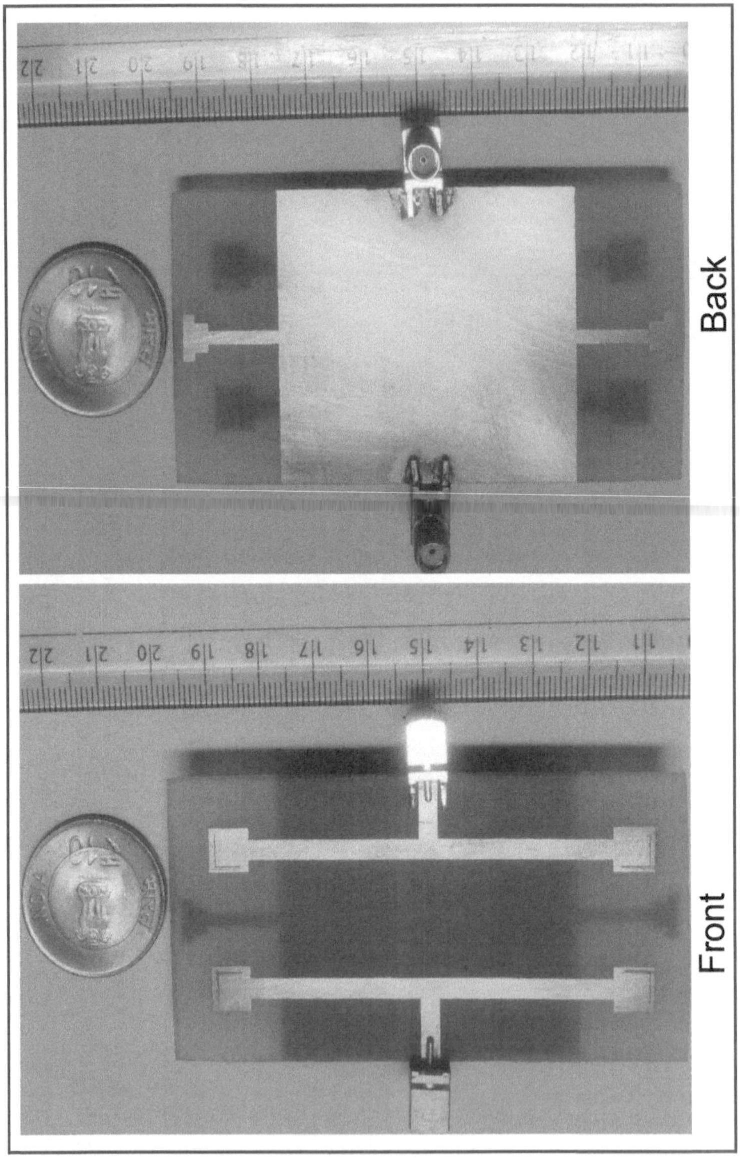

FIGURE 11.20
Fabricated views of 2×2 MIMO antenna [257]. [From: Leeladhar Malviya, M. V. Kartikeyan, and Rajib K. Panigrahi, "Multi-standard, multi-band planar MIMO antenna with diversity effects for wireless applications," International Journal of RF and Microwave Computer Aided Engineering, pp. 1–8, September 2018. Reproduced courtesy of the John Wiley & Sons, Ltd.]

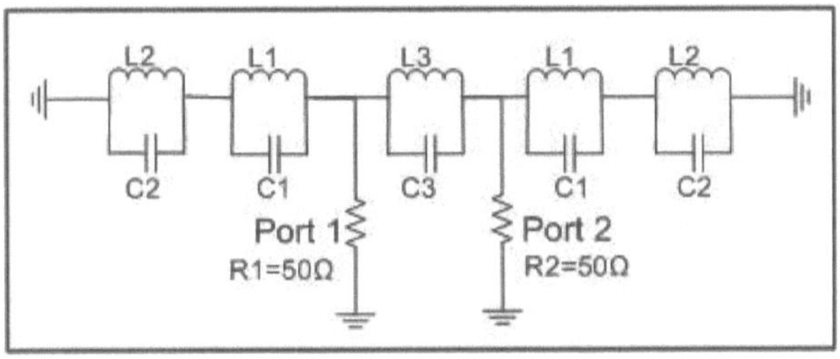

FIGURE 11.21
Equivalent circuit of compact MIMO antenna [257]. [From: Leeladhar Malviya, M. V. Kartikeyan, and Rajib K. Panigrahi, "Multi-standard, multi-band planar MIMO antenna with diversity effects for wireless applications," International Journal of RF and Microwave Computer Aided Engineering, pp. 1–8, September 2018. Reproduced courtesy of the John Wiley & Sons, Ltd.]

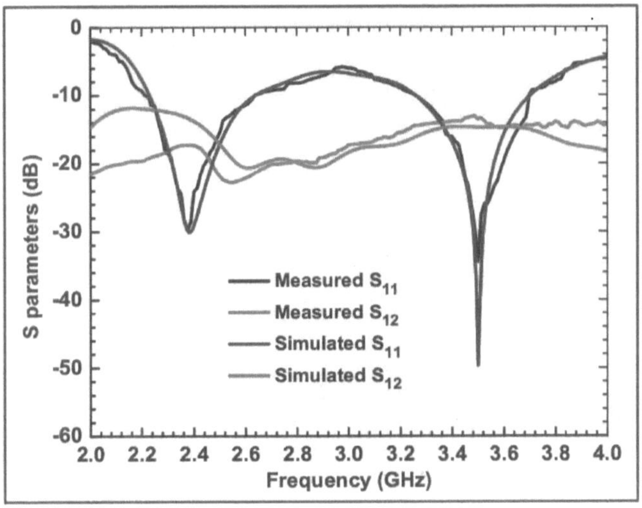

FIGURE 11.22
S parameters of 2 × 2 MIMO antenna [257]. [From: Leeladhar Malviya, M. V. Kartikeyan, and Rajib K. Panigrahi, "Multi-standard, multi-band planar MIMO antenna with diversity effects for wireless applications," International Journal of RF and Microwave Computer Aided Engineering, pp. 1–8, September 2018. Reproduced courtesy of the John Wiley & Sons, Ltd.]

Also, the bandwidth in each band is better than 400 MHz. Figure 11.22 shows the comparison of *S* parameters of the designed compact MIMO.

Surface current distributions of the 2×2 MIMO antenna are shown in Fig. 11.23 for the 2.38 GHz and 3.5 GHz frequencies. Effect of coupling can be observed using the surface current distribution. Due to symmetry, port 1 is

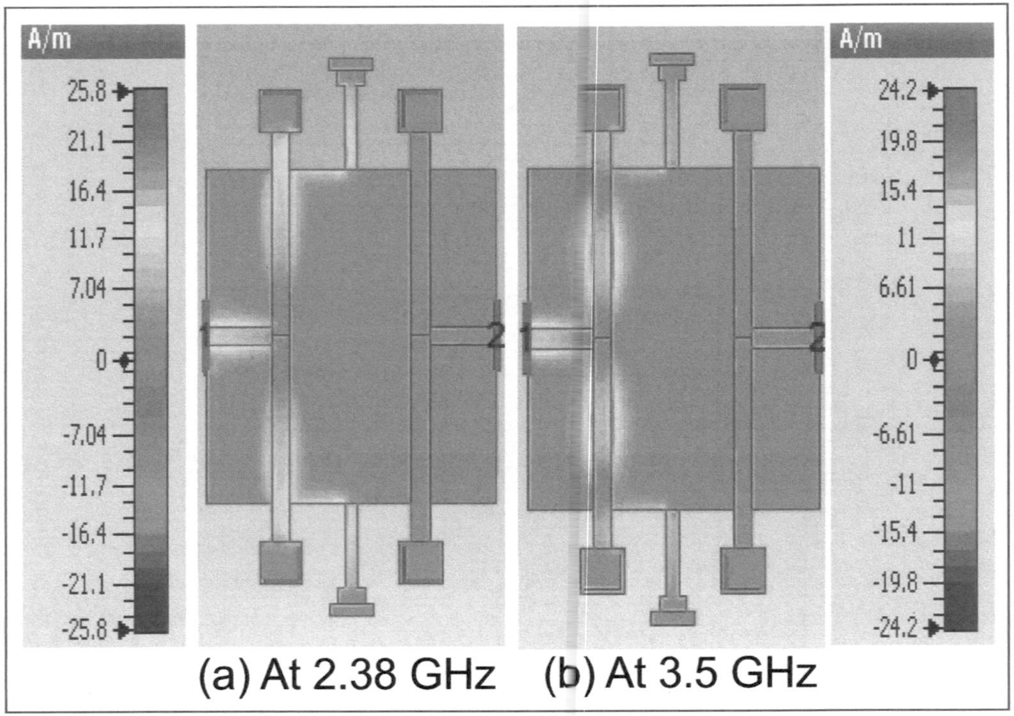

FIGURE 11.23
Current distribution of 2×2 MIMO antenna [257]. [From: Leeladhar Malviya, M. V. Kartikeyan, and Rajib K. Panigrahi, "Multi-standard, multi-band planar MIMO antenna with diversity effects for wireless applications," International Journal of RF and Microwave Computer Aided Engineering, pp. 1–8, September 2018. Reproduced courtesy of the John Wiley & Sons, Ltd.]

excited only and discussed for the current distribution. The same concept is considered when other or both the ports are excited. Absence of ground stub results in high current linking with the un-excited port. Whereas the presence of ground stub results in lower coupling with un-excited port, and is the cause of required isolation at 2.38/3.5 GHz frequencies. For 2.38 GHz, current can be observed on the ground stub. Whereas, at 3.5 GHz, current concentrates on the feed arms. In both the resonant cases, better than 12.5 dB isolation is achieved.

Far-field gain of the compact MIMO antenna is measured in an anechoic chamber to validate the CST-MWS gain response using the substitution method. Figures 11.24 and 11.25 show the simulated and measured MIMO antenna gain, which is more than 2.7 dBi in lower and higher bands. At resonant frequencies, the gain is more than 4.0 dBi. Similarly, simulated radiation efficiency in both the bands is more than 73%. Radiation efficiencies at 2.38 GHz and 3.5 GHz resonant frequencies are 96.23% and 87%.

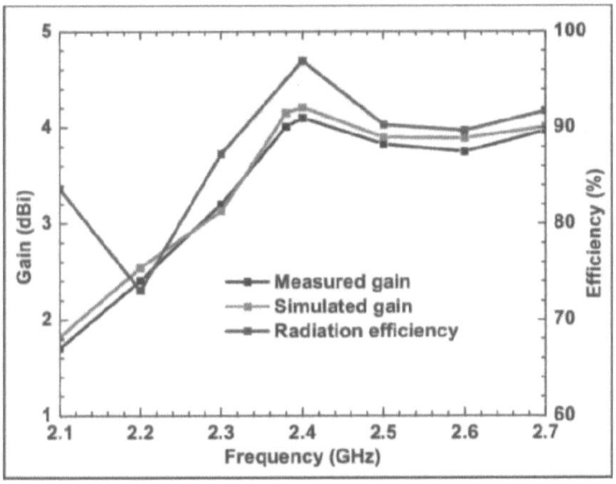

FIGURE 11.24

Gain and efficiency in lower band [257]. [From: Leeladhar Malviya, M. V. Kartikeyan, and Rajib K. Panigrahi, "Multi-standard, multi-band planar MIMO antenna with diversity effects for wireless applications," International Journal of RF and Microwave Computer Aided Engineering, pp. 1–8, September 2018. Reproduced courtesy of the John Wiley & Sons, Ltd.]

Diversity performance in terms of ECC is obtained using (4.7). For both the lower and upper bands, ECC is in the range of 10^{-3}. Figure 11.26 shows the simulated and measured values of the ECC. Variation in measured ECC is due to the fabrication errors and port coupling losses.

Similarly, TARC of the presented MIMO may be obtained using (4.10). Value of TARC is zero for no reflection, and one for total reflection. Figure 11.27 shows, different TARC curves for different combinations of phase excitations at two ports. The best combination is obtained for excitation angle

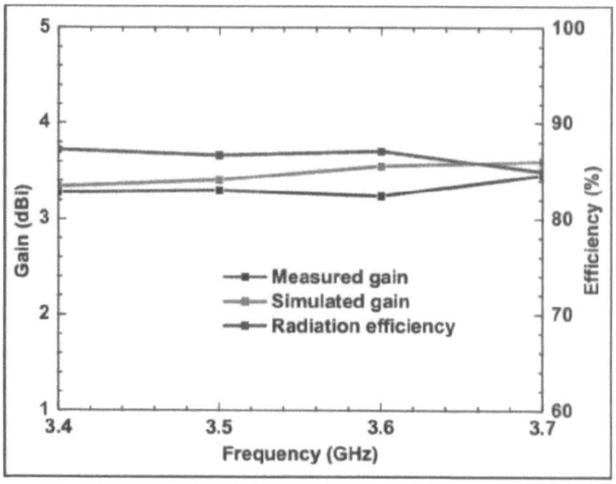

FIGURE 11.25
Gain and efficiency in higher band [257]. [From: Leeladhar Malviya, M. V. Kartikeyan, and Rajib K. Panigrahi, "Multi-standard, multi-band planar MIMO antenna with diversity effects for wireless applications," International Journal of RF and Microwave Computer Aided Engineering, pp. 1–8, September 2018. Reproduced courtesy of the John Wiley & Sons, Ltd.]

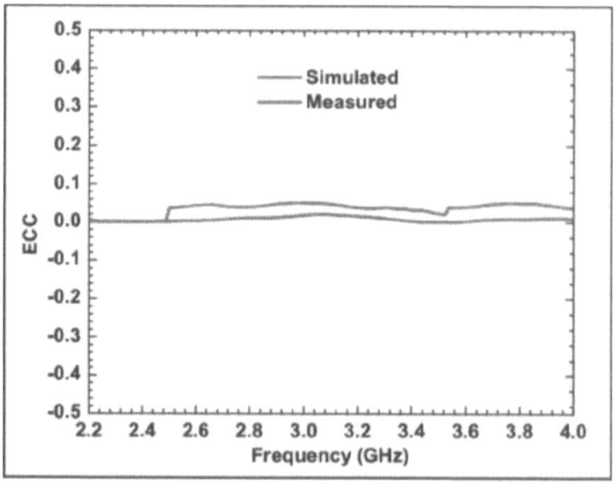

FIGURE 11.26
ECC of compact MIMO antenna [257]. [From: Leeladhar Malviya, M. V. Kartikeyan, and Rajib K. Panigrahi, "Multi-standard, multi-band planar MIMO antenna with diversity effects for wireless applications," International Journal of RF and Microwave Computer Aided Engineering, pp. 1–8, September 2018. Reproduced courtesy of the John Wiley & Sons, Ltd.]

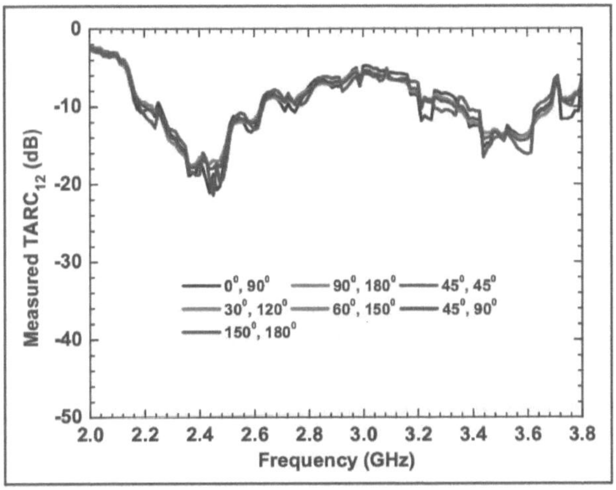

FIGURE 11.27
TARC of compact MIMO antenna [257]. [From: Leeladhar Malviya, M. V. Kartikeyan, and Rajib K. Panigrahi, "Multi-standard, multi-band planar MIMO antenna with diversity effects for wireless applications," International Journal of RF and Microwave Computer Aided Engineering, pp. 1–8, September 2018. Reproduced courtesy of the John Wiley & Sons, Ltd.]

combination of 30°, 120°, and has active bandwidth of 400 MHz in both the frequency bands. Impact of noise can be observed here on TARC responses. Most of the curves of TARC have similar responses for 2:1 VSWR bands.

Diversity performance in terms of MEG is obtained using (4.15). Let horizontal and vertical components of the Gaussian signals have mean $(\mu) = 0$ and variance $(\sigma) = 20$, respectively. Figure 11.28 compares MEG responses for outdoor environment ($XPR = 0$ dB) and indoor environment ($XPR = 6$ dB). MEG for isotropic and Gaussian environments have <-3 dB for both the bands. Hence, the presented design may be applicable for both the indoor and outdoor environments.

The S parameter responses of CST-MWS and equivalent circuit are shown in Fig. 11.29 using ADS simulation. Observed responses have a frequency shift only, and are in good agreement.

CST-MWS far-field patterns are validated in an anechoic chamber for the validity of the design. Main lobe directions of E-field patterns for two radiators are 192.0° and 168.0° at 2.38 GHz, and 351.0° and 9.0° at 3.5 GHz resonant frequencies. Beam widths are 82.8° at 2.38 GHz and 69.6° at 3.5 GHz. Similarly, H-field main lobe direction is 324.0° at 2.38 GHz, and 214.0° at 3.5 GHz for two ports. Beam widths of H-field for two radiators for 2.38 GHz is 134.4°, and at 3.5 GHz is 57.0°. These patterns have very small differences, and are in good agreement. Figures 11.30, 11.31, 11.32, and 11.33 show the

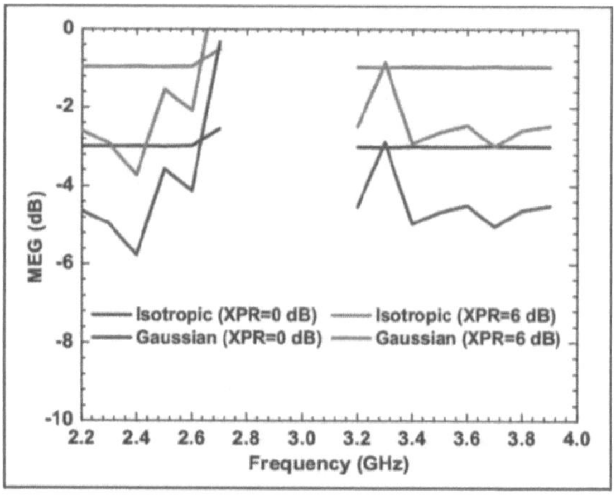

FIGURE 11.28
MEG of compact MIMO antenna [257]. [From: Leeladhar Malviya, M. V. Kartikeyan, and Rajib K. Panigrahi, "Multi-standard, multi-band planar MIMO antenna with diversity effects for wireless applications," International Journal of RF and Microwave Computer Aided Engineering, pp. 1–8, September 2018. Reproduced courtesy of the John Wiley & Sons, Ltd.]

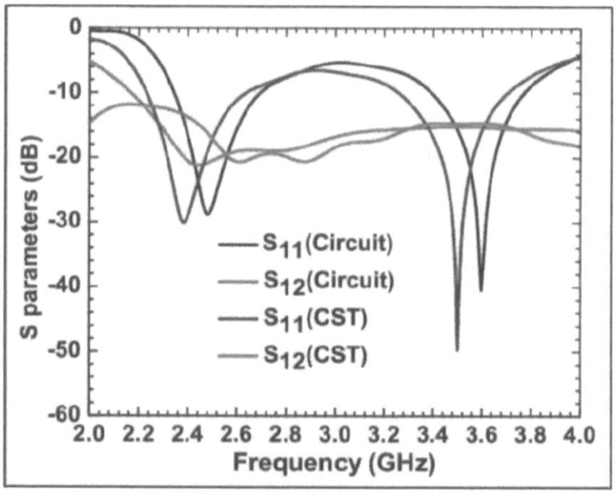

FIGURE 11.29
Circuit and CST *S* parameters of MIMO antenna [257]. [From: Leeladhar Malviya, M. V. Kartikeyan, and Rajib K. Panigrahi, "Multi-standard, multi-band planar MIMO antenna with diversity effects for wireless applications," International Journal of RF and Microwave Computer Aided Engineering, pp. 1–8, September 2018. Reproduced courtesy of the John Wiley & Sons, Ltd.]

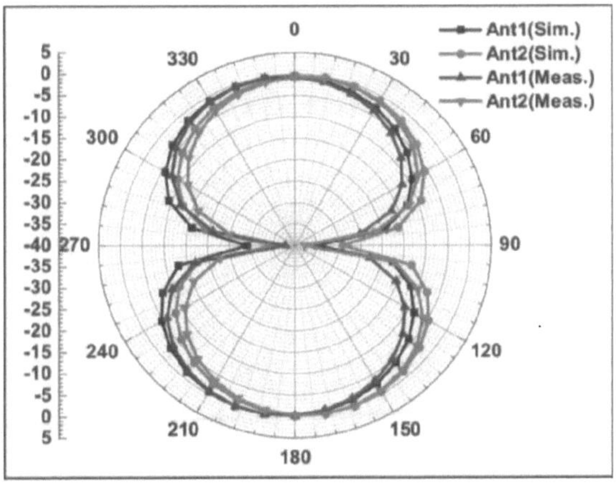

FIGURE 11.30
E-field radiation patterns at 2.38 GHz [257]. [From: Leeladhar Malviya, M. V. Kartikeyan, and Rajib K. Panigrahi, "Multi-standard, multi-band planar MIMO antenna with diversity effects for wireless applications," International Journal of RF and Microwave Computer Aided Engineering, pp. 1–8, September 2018. Reproduced courtesy of the John Wiley & Sons, Ltd.]

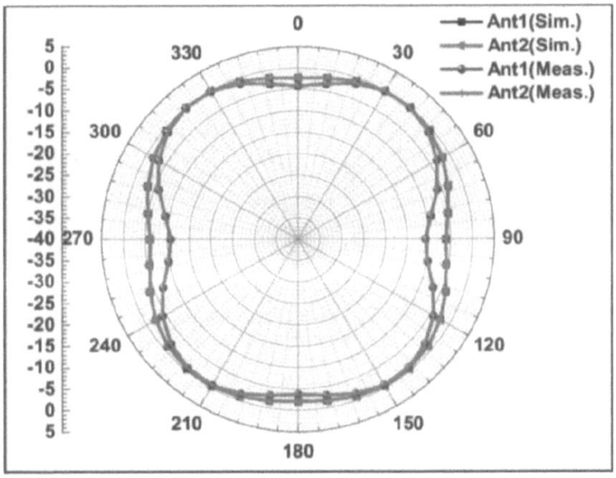

FIGURE 11.31
H-field radiation patterns at 2.38 GHz [257]. [From: Leeladhar Malviya, M. V. Kartikeyan, and Rajib K. Panigrahi, "Multi-standard, multi-band planar MIMO antenna with diversity effects for wireless applications," International Journal of RF and Microwave Computer Aided Engineering, pp. 1–8, September 2018. Reproduced courtesy of the John Wiley & Sons, Ltd.]

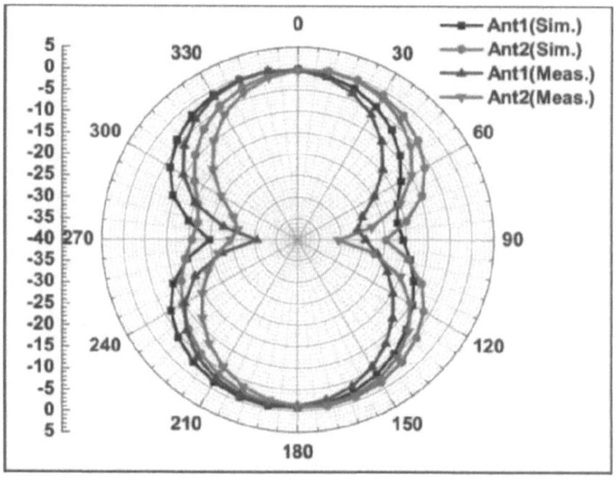

FIGURE 11.32
E-field radiation patterns at 3.5 GHz [257]. [From: Leeladhar Malviya, M. V. Kartikeyan, and Rajib K. Panigrahi, "Multi-standard, multi-band planar MIMO antenna with diversity effects for wireless applications," International Journal of RF and Microwave Computer Aided Engineering, pp. 1–8, September 2018. Reproduced courtesy of the John Wiley & Sons, Ltd.]

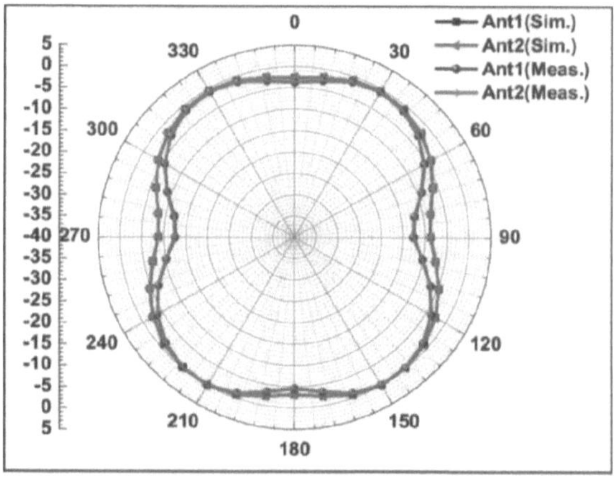

FIGURE 11.33
H-field radiation patterns at 3.5 GHz [257]. [From: Leeladhar Malviya, M. V. Kartikeyan, and Rajib K. Panigrahi, "Multi-standard, multi-band planar MIMO antenna with diversity effects for wireless applications," International Journal of RF and Microwave Computer Aided Engineering, pp. 1–8, September 2018. Reproduced courtesy of the John Wiley & Sons, Ltd.]

comparison of CST-MWS and measured far-field patterns for the presented compact MIMO antenna.

TABLE 11.4
Comparison of compact MIMO with earlier reported works [257] [From: Leeladhar Malviya, M. V. Kartikeyan, and Rajib K. Panigrahi, "Multi-standard, multi-band planar MIMO antenna with diversity effects for wireless applications," International Journal of RF and Microwave Computer Aided Engineering, pp. 1–8, September 2018. Reproduced courtesy of the John Wiley & Sons, Ltd.]

Ref. No.	Frequency/ bands (GHz)	No. of elements	Size (mm^2)	Isolation (dB)	ECC	Gain (dBi)
[94]	2.45	4	100 × 55	>10	<0.01	2.15
[120]	2.4	2	100 × 100	>40	—	3.4
[199]	2.4	4	100 × 60	>11	<0.3	>3.12
[200]	2.43–2.54	2	60 × 57	>30	$<10^{-3}$	>2.57
[254]	2.45	2	100 × 150	>15	$<10^{-3}$	> −0.7
[256]	2.24–2.64 3.41–3.69	4	65.3 × 65.3	>12.4	$<10^{-2}$	>2.6
In this design	2.24–2.64 3.26–3.70	4	45.1 × 90.2	>12.5	$<10^{-3}$	>2.7

Table 11.4 shows the comparison of the presented MIMO design with the available reported works. Designed compact MIMO has an excellent performance in terms of bandwidth, ECC, and less dimension, with earlier reported works. Reference [120] has good isolation but dimension is very big, single band, and has two elements only. Similarly, reference [199] has a single band and a large area of implementation for four elements. Hence, the designed compact MIMO has much better performance in terms of the number of elements, size, and bandwidth in comparison with earlier reported works. The designed MIMO can be extended to make other designs also [255].

11.4 Concluding Remarks

In the first design, MIMO with pentagon-shaped patches, T-shaped isolator, and with diversity effects has been presented for 2.4/2.5/3.5 GHz WLAN/WiMAX applications with power divider arms. The presented MIMO exhibited more than 2.6 dBi gain in both the frequency bands and has more than 12.4 dB isolation at ports. The effect of equivalent circuit modeling explored its tunable property. The de-correlation property of antenna has been exhibited by the ECC, and has very less value than the limit set by ITU. Also, the presented MIMO has MEG as per the indoor and outdoor activities for isotropic and Gaussian mediums [256].

In the second design, the compact MIMO antenna design with diversity effects has been presented with power divider arms. The effect of isolation was included in terms of surface current and ECC for the designed MIMO. Effect of ground stub resulted in the lower mutual coupling between ports. Effect of equivalent circuit showed the tunability of the design. Designed MIMO can be used for 2.23–2.64 GHz and 3.26–3.70 GHz frequency bands, and is applicable for both the indoor and outdoor (isotropic and Gaussian) environments. A very low value of ECC showed the applicability of MIMO for wireless applications under set ITU standards [257].

12

Concluding Remarks and Future Perspective

In this book, microstrip patch-based 2×2 MIMO antennas with diversity effects have been discussed and examined. Study of different mutual coupling reduction techniques and diversity techniques have also been discussed to aware the designers and researchers to be familiar with it. The complete book is focused on the design of MIMO antennas (linearly and circularly polarized) with different types of ground structures like PSG, PEG, and offset to control the effects of mutual coupling. All the designs have discussed here with all the possible aspects of diversity like ECC, TARC, MEG, far-field effects like gain, efficiency, and radiation patterns etc.

12.1 Contribution of the Book

The detailed contributions of the book are as follows:

- An introduction of the entire MIMO antenna design work followed by the motivation behind it, research objectives, problem statement, and organization of the book is presented in Chapter 1.
- Wireless channel fading like large scale and small scale, and different interferences like CCI, ACI are discussed in detail. As well as approaches of capacity enhancement have also been discussed in depth in Chapter 2.
- MIMO functions, different types of MIMO, and applications of MIMO in data extension, range extension, bandwidth saving, power saving, etc., have been covered in Chapter 3.
- MIMO antenna design performance parameters like reflection coefficient, VSWR, transmit and reflect powers, transmission coefficient, isolation, ECC, TARC, and MEG are discussed in depth and covered in Chapter 4.
- Massive MIMO antenna technology with 5G frequency bands, spatial multiplexing, and different beamforming techniques have been discussed deeply and covered in Chapter 5.

- Chapter 6 presented an in-depth literature review of MIMO antennas with diversity and mutual coupling reduction techniques for beginners and researchers, who require an extensive review of the subject with the blueprint to trigger a pioneering design. The merits and demerits of mutual coupling reduction and diversity techniques of MIMO antennas have been presented and also discussed here thoroughly.

- Chapter 7 described introduction and related work of 2.4/5.2 GHz multi-band MIMO antenna. The implementations of multi-band folded MIMO antenna with diversity and partially stepped ground is covered here. All the possible results of diversity and far-field characterization are included for practical environmental conditions. The presented design achieved ECC $\leq$0.01 in each frequency band, which is lower than the set value by ITU for wireless applications.

- Chapter 8 described introduction and related work of wide-band MIMO antenna, implementations of MIMO antenna with diversity and partially extended ground, and different performance parameters for the validity of presented design with practical environmental conditions. The value of ECC is $\leq$0.01 in the presented MIMO antenna.

- Chapter 9 described introduction and related work of 5.8 GHz CP-MIMO antenna for NLOS communication. The implementations of CP-MIMO antenna with LHCP and RHCP polarization and possible aspects of design parameters are presented here for wireless signal continuity and reliability.

- Chapter 10 presented two LTE-MIMO antenna designs for wireless communication. MIMO with LTE enhances data rate drastically. Folded loop-shaped radiating structure is presented in first MIMO design, and mathematically inspired patch shape is used in the design of the second MIMO here. The introduction and related works, implementations of LTE MIMO antenna designs, and simulation and experimental results of practical environmental conditions have been discussed in detail.

- Chapter 11 describes MIMO antenna designs with power divider arms for wireless applications. First MIMO antenna design with pentagon-shaped radiator and with inverted L-shaped slot, and second MIMO antenna design with compact radiator with inverted L-shaped slot have designed and discussed in details here.

- Finally, Chapter 12 concludes with the book contributions toward the MIMO antenna designs for modern wireless applications. Future perspectives are included here for great work to be done for wireless communication.

12.2 Future Perspective

The future extent of the work may be explored to give better insight into the applications of MIMO antennas with mutual coupling reduction and diversity techniques for wireless applications, and enlisted as follows:

- Wireless application with diversity and mutual coupling reduction techniques showed amazing promise for designing and performance enhancement of MIMO antennas. Some hidden behavior is still to be investigated.
- Despite the fact that the diversity techniques and mutual coupling reduction techniques have been studied in a variety of MIMO antenna designs by the designers, there are still some possibilities to explore their actual behavior on the microstrip components like directional couplers, isolators, combiners, etc, for 5G, 6G and THz wireless communications.
- Beamforming MIMO antennas have been considered as an essential part of future generations of antenna applications. The isolation among the radiating elements and miniaturization of radiating elements and overall size is very important requirement of massive MIMO technology and is to be investigated deeply.
- The unified outcome of MIMO antennas with microstrip filters require plenty of attention due to the involvement of different resonant properties. The far-field pattern behavior with the regenerative nature of microstrip filter is of interest in the present scenario due to their conversion properties at 5G, 6G and THz applications.
- Apart from the above discussed future scopes of MIMO, a very important work that needs to be investigated is still pending which is the effect of the radiations of these designs on living and non-living bodies i.e. specific absorption rate (SAR).

Bibliography

[1] D. M. Pozar, "Microstrip antennas," Proceedings of the IEEE, vol. 80, no. 1, pp. 79–91, January 1992.

[2] J. Q. Howell, "Microstrip antennas," IEEE Transactions on Antennas and Propagation, vol. 23, no. 1, pp. 90–93, January 1975.

[3] C. A. Balanis, "Antenna theory and design," John Wiley and Sons publication, 2nd Edition, 2007.

[4] D. M. Pozar, "Microwave engineering," John Wiley and Sons publication, 3rd Edition, 2009.

[5] H. Prasad, and M. J. Akhtar, "CPW fed triple band antenna for sensor application," IEEE MTT-S International Microwave and RF Conference (IMaRC), pp. 333–336, December 2015.

[6] S. Awasthi, A. Biswas, and M. J. Akhtar, "Dual-band dielectric resonator bandstop filters," International Journal of RF and Microwave Computer-Aided Engineering, vol. 25, no. 4, pp. 282–288, May 2015.

[7] A. K. Jha, and M. J. Akhtar, "An improved rectangular cavity approach for measurement of complex permeability of materials," IEEE Transactions on Instrumentation and Measurement, vol. 64, no. 4, pp. 995–1003, April 2015.

[8] S. Panda, N. K. Tiwari, and M. J. Akhtar, "Computationally intelligent sensor system for microwave characterization of dielectric sheets," IEEE Sensors Journal, vol. 16, no. 20, pp. 7483–7493, October 2016.

[9] G. A. Mavridis, D. E. Anagnostou, and M. T. Chryssomallis, "Evaluation of the quality factor, Q, of electrically small microstrip patch antennas," IEEE Antennas and Propagation Magazine, vol. 53, no. 4, pp. 216–224, August 2011.

[10] M. A. Al Tarifi, D. E. Anagnostou, A. Amert, and K. W. Whites, "Bandwidth enhancement of the resonant cavity antenna by using two dielectric superstrates," IEEE Transactions on Antennas and Propagation, vol. 61, no. 4, pp. 1898–1908, April 2013.

[11] X. Quan, R. Li, F. Yi, and D. E. Anagnostou, "Analysis and design of a 45° slant polarized omnidirectional antenna," IEEE Transactions on Antennas and Propagation, vol. 62, no. 1, pp. 86–93, January 2014.

[12] G. Kumar, and K. P. Ray, "Broadband microstrip antennas," Artech House, Norwood, USA, pp. 1–8, 2003.

[13] K. R. Carver, and J. W. Mink, "Microstrip antenna technology," IEEE Transactions on Antennas and Propagation, vol. 29, no. 1, pp. 2–24, January 1981.

[14] J. R. James, and P. S. Hall, "Handbook of microstrip antennas," Short Run Press Ltd, London, vol. 2, pp. 1079–1151, 1989.

[15] C. A. Balanis, "Advanced engineering electromagnetics," John Wiley and Sons publication, 2nd Edition, 2012.

[16] A. J. Paulraj, D. A. Gore, R. U. Nabar, and H. Bolcskei, "An overview of MIMO communications-A key to Gibabit wireless," Proceedings of the IEEE, vol. 92, no. 2, pp. 198–218, February 2004.

[17] S. Dey, and N. C. Karmakar, "Design of novel super wide band antennas close to the small antenna limitation theory," IEEE MTT-S International Microwave Symposium (IMS2014), pp. 1–4, June 2014.

[18] N. C. Karmakar, "Investigations into a cavity backed circular patch antenna," IEEE Transactions on Antennas and Propagation, vol. 50, no. 12, pp. 1706–1715, December 2002.

[19] N. C. Karmakar, P. Hendro, and L. S. Firmansyah, "Shorting strap tunable single feed dual-band PIFA," IEEE Microwave and Wireless Component Letters, vol. 13, no. 1, pp. 13–15, January 2003.

[20] Z. Ying, "Antennas in cellular phones for mobile communications," Proceedings of the IEEE, vol. 100, no. 7, pp. 2286–2296, July 2012.

[21] J. W. Wallace, J. B. Andersen, B. Daneshrad, B. K. Lau, and J. I. Takada, "Guest editorial for the special issue on multiple input multiple output," IEEE Transactions on Antennas and Propagation, vol. 60, no. 2, pp. 434–437, February 2012.

[22] M. A. Jensen, and J. W. Wallace, "A review of antennas and propagation for MIMO wireless communications," IEEE Transactions on Antennas and Propagation, vol. 52, no. 11, pp. 2810–2824, November 2004.

[23] C. R. Bocio, R. Gonzalo, M. S. Ayza, and M. Thumm, "Optimal horn antenna design to excite high order gaussian beam modes from TE smooth circular waveguide modes," IEEE Transactions on Antennas and Propagation, vol. 47, no. 9, pp. 1440–1448, September 1999.

[24] T. Lopetegi, M. A. G. Laso, M. J. Erro, M. Sorolla, and M. Thumm, "Analysis and design of periodic structures for microstrip lines by using the coupled mode theory," IEEE Microwave and Wireless Components Letters, vol. 12, no. 11, pp. 441–443, November 2002.

[25] G. P. Rao, K. Agarwal, M. V. Kartikeyan, M. K. Thumm, "Design of multiple beams forming network for switched beam antenna system with E shaped microstrip antenna," 31st International Conference on Infrared Millimeter Waves and 14th International Conference on Teraherz Electronics, pp. 439–439, September 2006.

[26] K. V. Klooster, M. Petelin, M. Thumm, and M. Aloisio, "Ka-band ground station antenna aspects for deep space telecommunication and radar," 7th European Conference on Antennas and Propagation (EuCAP), pp. 242–246, April 2013.

[27] M. Thottappan, and P. K. Jain, "Analysis of two dimensional metal electromagnetic band gap (EBG) structure using finite difference time domain method," IEEE International Vacuum Electronics Conference (IVEC), pp. 311–312, February 2011.

[28] S. W. Ellingson, "Antenna design and site planning considerations for MIMO," IEEE Vehicular Technology Conference, vol. 3, pp. 1718–1722, September 2005.

[29] K. C. J. Raju, "Experimental study on LTCC glass ceramic based dual segment cylindrical dielectric resonator antenna," Journal of Ceramics, Hindawi Publishing Corporation, vol. 2013, March 2013.

[30] K. Na, H. Ma, J. Park, J. Yeo, J. U. Park, and F. Bien, "Graphene based wireless environmental gas sensor on PET substrate," IEEE Sensors Journal, vol. 16, no. 12, pp. 5003–5009, June 2016.

[31] Z. Liu, and F. Bien, "A new multi-target detection for vehicle radar," International SoC Design Conference (ISOCC), pp. 28–29, November 2014.

[32] Y. S. Shin, J. H. Choi, B. N. Kim, and S. O. Park, "A monopole antenna with a magneto dielectric material and MIMO applications," IEEE Antennas and Wireless Propagation Letters, vol. 7, pp. 764–768, November 2008.

[33] J. Schiller, "Mobile communication," Pearson Education Ltd, England, 2nd Edition, 2003.

[34] T. S. Rappaport, "Wireless communications: Principles and practice," Prentice Hall Ltd, 2nd Edition, 2002.

[35] A. F. Molisch, "Wireless communications," Wiley Publication Ltd, 2nd Edition, 2010.

[36] V. K. Garg, and J. E. Wilkes, "Principles and applications of GSM," Prentice Hall Ltd, 2nd Edition, 1999.

[37] Noha Hassan and Xavier Fernando, "Massive MIMO Wireless Networks: An Overview," Electronics, vol. 63, no. 6, pp. 1–29, 2017.

[38] T. L. Marzetta, "Massive MIMO: An Introduction," Bell Labs Tech. J., vol. 20, pp. 11–20, 2015.

[39] C. E. Shannon, "A mathematical theory of communication," Bell System Technical Journal, vol. 27, pp. 379-423, pp. 623–656, October 1948.

[40] A. R. Kaye, and D. George, "Transmission of multiplexed PAM signals over multiple channel and diversity systems," IEEE Transactions on Communication Technology, vol. 18, no. 5, pp. 520–526, October 1970.

[41] L. H. Branderburg, and A. D. Wyner, "Capacity of the Gaussian channel with memory: The multivariate case," Bell System Technology Journal, vol. 53, no. 5, pp. 745–778, May–June 1974.

[42] W. Van Etten, "An optimum linear receiver for multiple channel digital transmission systems," IEEE Transactions on Communication, pp. 828–834, August 1975.

[43] W. Van Etten, "Maximum likelihood receiver for multiple channel transmission systems," IEEE Transactions on Communication, pp. 276–283, February 1976.

[44] J. H. Winters, "Optimum combining in digital mobile radio with co-channel interference," IEEE Journals on Selected Areas in Communications, vol. 2. pp. 528–539, July 1984.

[45] J. Salz, "Digital transmission over cross coupled linear channels," Technical Journal AT & T, vol. 64, no. 6, pp. 1147–1159, July–August 1985.

[46] R. Roy, and O. Bjorn, "Spatial division multiple access wireless communication systems," U.S. Patent no. 5515378, May 1996.

[47] G. G. Rayleigh, and J. M. Cioffi, "Spatio-temporal coding for wireless communication," IEEE Transactions on Communications, vol. 46, no. 3, pp. 357–366, March 1998.

[48] G. J. Foschini, "Layered space time architecture for wireless communication in a fading environment when using multiple antennas," Bell Labs Systems Technology, no. 1, pp. 41–59, August 1996.

[49] E. Telatar, "Capacity of multi antenna Gaussian channels," European Transactions on Telecommunications, vol. 10, no. 6, pp. 585–595, November–December 1999.

[50] A. Paulraj, and T. Kailath, "Increasing capacity in wireless braodcast systems using distributed transmission/directional reception," U. S. Patent no. 5345599, September 1994.

[51] R. T. Becker, "Pre-coding and spatially multiplexed MIMO in 3GPP long term evolution," Agilent Technologies, High frequency Electronics, pp. 18–26, October 2009.

[52] M. Costa, "Writing on dirty paper," IEEE Transactions on Information Theory, vol. 29, no. 3, pp. 439–441, May 1983.

[53] S. Cui, A. J. Goldsmith, and A. Bahai, "Energy efficiency of MIMO and cooperative MIMO techniques in sensor networks," IEEE Journal on selected areas in Communications, vol. 22, no. 6, pp. 1089–1098, August 2004.

[54] D. A. Basnayaka, P. J. Smith, and P. A. Martin, "Performance analysis of macrodiversity MIMO systems with MMSE and ZF receivers in flat rayleigh fading," IEEE transactions on wireless communications, vol. 12, no. 5, pp. 2240–2251, May 2013.

[55] J. Hoydis, K. Hosseini, S. Ten Brink, and M. Debbah, "Making smart use of excess antennas: Massive MIMO, small cells, and TDD," Bell Laboratory Technical Journal, vol. 18, no. 2, pp. 5–21, September 2013.

[56] S. H. Chae, S. K. Oh, and S. O. Park, "Analysis of mutual coupling, correlations, and TARC in WiBro MIMO array antenna," IEEE Antennas and Wireless Propagation Letters, vol. 7, pp. 122–125, February 2007.

[57] H. S. Singh, G. K. Pandey, P. K. Bharti, and M. K. Meshram, "A compact dual-band diversity antenna for WLAN applications with high isolation," Microwave and Optical Technology Letters, vol. 57, no. 4, pp. 906–912, April 2015.

[58] D. W. Bliss, and S. Govindasamy, "Adaptive wireless communications," Cambridge University Press, first edition, 2013.

[59] E. Dahlman, S. Parkvall, and J. Skold, "4G: LTE/LTE-advanced for mobile broadband," Elsevier, second edition, 2013.

[60] E. Biglieri, R. Calderbank, A. Constantinides, A. Goldsmith, A. Paulraj, and H. Poor, "MIMO wireless communications," Cambridge University Press, first edition, 2010.

[61] J. Hampton, "Introduction to MIMO communications," Cambridge University Press, first edition, 2013.

[62] A. Moradikordalivand, C. Y. Leow, T. A. Rahman, S. Ebrahimi, and T. H. Chua, "Wideband MIMO antenna system with dual polarization for Wi-Fi and LTE applications," International Journal of Microwave and Wireless Technologies, vol. 8, no. 3, pp. 643–650, 2016.

[63] Y. Li, C. Wang, H. Yuan, N. Liu, H. Zhao, and X. Li, "A 5G MIMO antenna manufactured by 3D printing method," IEEE Microwave and Wireless Components Lett., vol. 16, pp. 657–660, 2016.

[64] M. R. Ramli, S. K. A. Rahim, H. A. Rahman, M. I. Sabran, and M. L. Samingan, "Flexible microstrip grid array polymer-conductive rubber antenna for 5G mobile communication application," Wiley Online Library, vol. 59, no. 8, pp. 1866–1870, 2017.

[65] Xudong Cheng, Yejun He, Li Zhang, and Jian Qiao, "Channel modeling and analysis for multipolarized massive MIMO systems," International Journal of Communication Systems, pp. 1–16, April 2018.

[66] Emil Björnson, Erik G. Larsson, and Thomas L. Marzetta, "Massive MIMO:Ten myths and one critical question," IEEE Communications Magazine, vol. 59, no. 8, pp. 114–123, 2016.

[67] Ehab Ali, Mahamod Ismail, Rosdiadee Nordin, and Nor Fadzilah Abdullah, "Review: Beamforming techniques for massive MIMO systems in 5G:overview, classification, and trends for future research," Front Inform Technol Electron Eng, vol. 18, no. 6, pp. 753–772, 2017.

[68] W. Liu, and S. Weiss, "Wideband beamforming: Concepts and Techniques," John Wiley & Sons, UK, 2010.

[69] N. Tiwari, and T. R. Rao, "A switched beam antenna array with Butler matrix network using substrate integrated waveguide technology for 60 GHz communications," Int. Conf. on Advances in Computing, Communications and Informatics, pp. 2152–2157, 2015.

[70] L. Huang, B. Zhang, and Z. F. Ye, "Robust adaptive beamforming using a new projection approach," IEEE Int. Conf. on Digital Signal Processing, pp. 1181–1185, 2015.

[71] M. Z. A. Bhotto, and I. V. Bajić, "Constant modulus blind adaptive beamforming based on unscented Kalman filtering," IEEE Signal Process. Lett., vol. 22, no. 4, pp. 474–478, 2015.

[72] A. Arunitha, T. Gunasekaran, and N. S. Kumar, "Adaptive beam forming algorithms for MIMO antenna," Int. J. Innov. Technol. Explor. Eng., vol. 14, no. 8, pp. 9–12, 2015.

[73] S. Darzi, T. S. Kiong, and M. T. Islam, "A memory based gravitational search algorithm for enhancing minimum variance distortionless

response beamforming," Applied Soft Computing, vol. 47, pp. 103–118, 2016.

[74] S. Noh, M. D. Zoltowski, and D. J. Love, "Training sequence design for feedback assisted hybrid beamforming in massive MIMO systems," IEEE Transaction on Communication, vol. 64, no. 1, pp. 187–200, 2016.

[75] R. G. Vaughan, and J. B. Anderson, "Antenna diversity in mobile communications," IEEE Transactions on Vehicle Technology, vol. 36, no. 4, pp. 149–172, November 1987.

[76] C. R. Paul, "Introduction to electro-magnetic compatibility," Wiley & Sons, Inc. Publication, Edition 2/e, pp. 62–78, March 1992.

[77] M. P. Karaboikis, V. C. Papamichael, G. F. Tsachtsiris, C. F. Soras, and V. T. Makios, "Integrating compact printed antennas onto small diversity/MIMO terminals," IEEE Transactions on Antennas and Propagation, vol. 56, no. 7, pp. 2067–2078, July 2008.

[78] S. Y. Lin, and I. H. Liu, "Small inverted U loop antenna for MIMO applications," Progress in Electromagnetics Research C, vol. 34, pp. 69–84, January 2013.

[79] G. T. Jeong, S. Choi, K. T. Lee, and W. Su Kim, "Low profile dual-wideband MIMO antenna with low ECC for LTE and Wi-Fi applications," International Journal of Antennas and Propagation, research article, vol. 2014, pp. 1–6, May 2014.

[80] M. U. Khan, and M. S. Sharawi, "A dual-band microstrip annular slot based MIMO antenna system," Microwave and Optical Technology Letters, vol. 57, no. 2, pp. 360–364, February 2015.

[81] W. C. Y. Lee, and Y. U. S. Yeh, "Polarization diversity system for mobile radio," IEEE Transactions on Communications, vol. 20, no. 5, pp. 912–923, October 1972.

[82] S. B. Yeap, X. Chen, J. A. Dupuy, C. C. Chiau, and C. G. Parini, "Integrated diversity antenna for laptop and PDA terminal in a MIMO system," IEE proceedings on Microwave Antennas and Propagation, vol. 152, no. 6, pp. 495–504, December 2005.

[83] J. Oh, and K. Sarabandi, "Compact, low profile, common aperture polarization, and pattern diversity antennas," IEEE Transactions on Antennas and Propagation, vol. 62, no. 2, pp. 569–576, February 2014.

[84] H. Zhang, Z. Wang, J. Yu, and J. Huang, "A compact MIMO antenna for wireless communication," IEEE Antennas and Propagation Magazine, vol. 50, no. 6, pp. 104–107, December 2008.

[85] Aicha Mchbal, Naima Amar Touhmani, Hanae Elftouh, Mahmoud Moubadir, and Aziz Dkiouak, "Spatial and polarization diversity performance analysis of a compact MIMO antenna," 12th International Conference Interdisciplinarity in Engineering, Science Direct, Elsevier, no. 32, pp. 647–652, 2019.

[86] G. Adamiuk, S. Beer, W. Wiesbeck, and T. Zwick, "Dual orthogonal polarization antenna for UWR-IR technology," IEEE Antennas and Wireless Propagation Letters, vol. 8, pp. 981–984, August 2009.

[87] N. S. Awang Da, M. R. Dzulkifli, and M. R. Kamarudin, "Polarization diversity monopole antenna," PIERS proceedings, Cambridge, USA, pp. 466–469, July 2010.

[88] S. Zhang, P. Zetterberg, and S. He, "Printed MIMO antenna system of four closely spaced elements with large bandwidth and high isolation," Electronics Letters, vol. 46, no. 15, pp. 1052–1053, July 2010.

[89] M. Han, and J. Choi, "Dual-band MIMO antenna using polarization diversity for 4G mobile handset application," Microwave and Optical Technology Letters, vol. 53, no. 9, pp. 2075–2078, September 2011.

[90] J. R. Costa, E. B. Lima, C. R. Medeiros, and C. A. Fernandes, "Evaluation of a new wide-band slot array for MIMO performance enhancement in indoor WLANs," IEEE Transactions on Antennas and Propagation, vol. 59, no. 4, pp. 1200–1206, April 2011.

[91] S. W. Su, and C. T. Lee, "Low cost dual-loop antenna system for dual WLAN band access points," IEEE Transactions on Antennas and Propagation, vol. 59, no. 5, pp. 1652–1659, May 2011.

[92] J. Xiong, M. Zhao, H. Li, Z. Ying, and B. Wang, "Collocated electric and magnetic dipoles with extremely low correlation as a reference antenna for polarization diversity MIMO applications," IEEE Antennas and Wireless Propagation Letters, vol. 11, pp. 423–426, April 2012.

[93] H. Wi, B. Kim, W. Jang, and B. Lee, "Multiband handset antenna analysis including LTE band MIMO service," Progress in Electromagnetic Research, vol. 138, pp. 661–673, April 2013.

[94] S. Ghosh, T. N. Tran, and T. Le Ngoc, "Miniaturized four elements diversity PIFA," IEEE Antennas and Wireless Propagation Letters, vol. 12, pp. 396–400, March 2013.

[95] M. Koohestani, A. A. Moreira, and A. K. Skrivervik, "A novel compact CPW fed polarization diversity ultra-wideband antenna," IEEE Antennas and Wireless Propagation Letters, vol. 13, pp. 563–566, March 2014.

[96] A. Moradikorordalivand, T. A. Rahman, and M. Khalily, "Common elements wideband MIMO antenna system for WiFi/LTE access point applications," IEEE Antennas and Wireless Propagation Letters, vol. 13, pp. 1601–1604, July 2014.

[97] D. Piazza, N. J. Kirsch, A. Forenza, R. W. Heath, and K. R. Dandekar, "Design and evaluation of a reconfigurable antenna array for MIMO systems," IEEE Transactions on Antenna and Propagation, vol. 56, no. 3, pp. 869–881, March 2008.

[98] Y. Cai, and Z. Du, "A novel pattern reconfigurable antenna array for diversity systems," IEEE Antennas and Wireless Propagation Letters, vol. 8, pp. 1227–1230, November 2009.

[99] A. H. Ramadan, J. Constantine, Y. Tawk, C. G. Christodoulou, and K. Y. Kabalan, "Frequency tunable and pattern diversity antennas for cognitive radio applications," International Journal of Antenna and Propagation, Hindawi Publishing Corporation, vol. 2014, pp. 7–13, March 2014.

[100] C. H. See, R. A. Abd Alhameed, N. J. Mcewan, S. M. R. Jones, R. Asif, and P. S. Excell, "Design of a printed MIMO/Diversity monopole antenna for future generation handheld devices," International Journal of RF and Microwave Computer Aided Engineering, vol. 24, no. 3, pp. 348–359, May 2014.

[101] M. Ayatollahi, Q. Rao, and D. Wang, "A Compact high isolation and wide bandwidth antenna array for long term evolution wireless devices," IEEE Transactions on Antennas and Propagation, vol. 60, no. 10, pp. 4960–4963, October 2012.

[102] R. Addaci, A. Diallo, C. Luxey, P. Le Thuc, and R. Staraj, "Dual-band WLAN diversity antenna system with high port to port isolation," IEEE Antennas and Wireless Propagation Letters, vol. 11, pp. 244–247, February 2012.

[103] Kommana Vasu Babu, and Bhuma Anuradha, "Analysis of multi-band circle MIMO antenna design for C-band applications," Progress in Electromagnetics Research C, vol. 91, pp. 185–196, March 2019.

[104] Z. Li, Z. Du, M. Takahashi, K. Saito, and K. Ito, "Reducing mutual coupling of MIMO antennas with parasitic elements for mobile terminals," IEEE Transactions on Antennas and Propagation, vol. 60, no. 2, pp. 473–481, February 2012.

[105] K. Payandehjoo, and R. Abhari, "Investigation of parasitic elements for coupling reduction in multi-antenna handset devices," International Journal of RF and Microwave Computer Aided Engineering, vol. 24, pp. 1–10, January 2014.

[106] A. Diallo, C. Luxey, P. Le Thuc, R. Staraj, and G. Kossiavas, "Study and reduction of mutual coupling between two mobile Phone PIFAs operating in the DCS 1800 and UMT bands," IEEE Transactions on Antennas and Propagation, vol. 54, no. 11, pp. 3063–3073, November 2006.

[107] A. Diallo, C. Luxey, P. Le Thuc, R. Staraj, and G. Kossiavas, "Enhanced two antenna structures for universal mobile telecommunications system diversity terminals," IET Microwaves Antennas and Propagation, vol. 2, pp. 93–101, February 2008.

[108] S. Wen Su, C. Tse Lee, and Fa Shian Chang, "Printed MIMO antenna system using neutralization line technique for wireless USB dongle applications," IEEE Transactions on Antennas and Propagation, vol. 60, no. 2, pp. 456–463, February 2012.

[109] H. Wang, L. Liu, Z. Zhang, Y. Li, and Z. Feng, "Ultra compact three port MIMO antenna with high isolation and directional radiation patterns," IEEE Antennas and Wireless Propagation Letters, vol. 13, pp. 1545–1548, July 2014.

[110] M. Karaboikis, C. Soras, G. Tsachtsiris, and V. Makios, "Compact dual printed inverted F antenna diversity systems for portable wireless devices," IEEE Antennas and Wireless Propagation Letters, vol. 3, pp. 9–14, December 2004.

[111] Y. S. Shin, and S. O. Park, "Spatial diversity antenna for WLAN application," Microwave and Optical Technology Letters, vol. 49, no. 6, pp. 1290–1294, June 2007.

[112] R. Addaci, K. Haneda, A. Diallo, P. Le Thuc, C. Luxey, R. Staraj, and P. Nainikainen "Dual-band WLAN multiantenna system and diversity/MIMO performance evaluation," IEEE Transactions on Antennas and Propagation, vol. 62, no. 3, pp. 1409–1415, March 2014.

[113] C. Yuk Chiu, C. Ho Cheng, R. D. Murch, and C. R. Rowell, "Reduction of mutual coupling between closely packed antenna elements," IEEE Transactions on Antennas and Propagation, vol. 55, no. 6, pp. 1732–1738, June 2007.

[114] R. A. Bhatti, J. H. Choi, and S. O. Park, "Quad band MIMO antenna array for portable wireless communications terminals," IEEE Antennas and Wireless Propagation Letters, vol. 8, pp. 129–132, April 2009.

[115] H. Li, J. Xiong, and S. He, "A compact planar MIMO antenna system of four elements with similar radiation characteristics and isolation structure," IEEE Transactions on Antenna and Propagation, vol. 8, no. 11, pp. 1107–1110, October 2009.

[116] S. L. Zuo, Y. Z. Yin, W. J. Wu, Z. Y. Zhang, and J. Ma, "Investigations of reduction of mutual coupling between two planar monopoles using two $\lambda/4$ slots," Progress in Electromagnetics Research Letters, vol. 19, pp. 9–18, November 2010.

[117] C. R. Medeiros, E. B. Lima, J. R. Costa, and C. A. Fernandes, "Wideband slot antenna for WLAN access point," IEEE Antennas and Wireless Propagation Letters, vol. 9, pp. 79–82, March 2010.

[118] S. Dumali, C. Raitton, and D. L. Paul, "A slot antenna array with low mutual coupling for use on small mobile terminals," IEEE Transactions on Antennas and Propagation, vol. 59, no. 5, pp. 1512–1520, May 2011.

[119] M. Sonkki, and E. Salonen, "Low mutual coupling between monopole antennas by using two slots," IEEE Antennas and Wireless Propagation Letters, vol. 9, pp. 138–141, March 2010.

[120] K. Wei, Z. Zhang, W. Chen, and Z. Feng, "A novel hybrid fed path antenna with pattern diversity," IEEE Antennas and Wireless Propagation Letters, vol. 9, pp. 562–565, May 2010.

[121] X. Zhou, X. Quan, and R. Li, "A dual broadband MIMO antenna system for GSM/UMTS/LTE and WLAN handsets," IEEE Antennas and Propagation Letters, vol. 11, pp. 551–554, May 2012.

[122] J. park, J. choi, Ji Yong Park, and K. Sarabandi, "Study of a T shaped slot with a capacitor for high isolation between MIMO antennas," IEEE Antennas and Wireless Propagation Lettters, vol. 11, pp. 1541–1544, October 2012.

[123] Q. Zeng, Y. Yao, S. Liu, J. Yu, P. Xie, and X. Chen, "Tetra-band small size printed strip MIMO antenna for mobile handset application," International Journal of Antennas and Propagation, research article, vol. 2012, pp. 1–8, January 2012.

[124] J. Andersen, and H. Rasmussen, "Decoupling and de-scattering networks for antennas," IEEE Transactions on Antennas and Propagation, vol. 24, no. 6, pp. 841–846, November 1976.

[125] A. C. K. Mak, C. R. Rowell, and R. D. Murch, "Isolation enhancement between two closely packed antennas," IEEE Transactions on Antennas and Propagation, vol. 56, no. 11, pp. 3411–3419, November 2008.

[126] S. C. Chen, Y. S. Wang, and S. J. Chung, "A decoupling technique for increasing the port isolation between two strongly coupled antennas," IEEE Transactions on Antennas and Propagation, vol. 56, no. 12, pp. 3650–3658, December 2008.

[127] Q. Gong, Y. C. Jiao, and S. X. Gong, "Compact MIMO antennas using a ring hybrid for WLAN applications," Journal of Electromagnetic Waves and Applications, vol. 25, no. 2-3, pp. 431–441, 2011.

[128] S. Chu, S. X. Gong, Y. Liu, W. Jiang, and Y. Guan, "Compact and low coupled monopole antennas for MIMO system applications," Journal of Electromagnetic Waves and Applications, vol. 25, no. 5-6, pp. 703–712, 2011.

[129] S. Cui, Y. Liu, W. Jiang, S. X. Gong, Y. Guan, and S. T. Yu, "A novel compact dual-band MIMO antenna with high port isolation," Journal of Electromagnetic Waves and Applications, vol. 25, no. 11-12, pp. 1645–1655, 2011.

[130] V. Ssorin, A. Artemenko, A. Sevastyanov, and R. Maslennikov, "Compact bandwidth optimized two element MIMO antenna system for 2.5-2.7 GHz band," Antenna and Propagation (EUCAP-2011), 5th European Conference, pp. 319–323, April 2011.

[131] S. Farsi, H. Aliakbarian, D. Schreurs, B. Nauwelaers, and G. A. E. Vanderbosch, "Mutual coupling reduction between planar antennas by using a simple microstrip U section," IEEE Antennas and Wireless Propagation Letters, vol. 11, pp. 1501–1503, December 2012.

[132] C. H. Wu, G. T. Zhou, Y. L. Wu, and T. G. Ma, "Stub loaded reactive decoupling network for two element array using even odd analysis," IEEE Antennas and Wireless Propagation Letters, vol. 12, pp. 452–455, March 2013.

[133] M. A. Moharram, and A. A. Kishk, "General decoupling network design between two coupled antennas for MIMO applications," Progress in Electromagnetics Research Letters, vol. 37, pp. 133–142, January 2013.

[134] M. Pelosi, M. B. Knudsen, and G. F. Pedersen,"Multiple antenna systems with inherently decoupled radiators," IEEE Transactions on Antennas and Propagation, vol. 60, no. 2, pp. 503–515, February 2012.

[135] K. Wang, R. A. M. Mauermayer, and T. F. Eibert, "Compact two element printer monopole array with partially extended ground plane," IEEE Antennas and Wireless Propagation Letters, vol. 13, pp. 138–140, January 2014.

[136] H. wang, Z. Zhang, and Z. Feng, "Dual port planar MIMO antenna with ultra high isolation and orthogonal radiation patterns," Electronics Letters, vol. 51, no. 1, pp. 7–8, January 2015.

[137] F. Yang, and Y. R. Samii, "Microstrip antennas integrated with electromagnetic bandgap (EBG) structures: A low mutual coupling design for

array applications," IEEE Transactions on Antennas and Propagation, 51, no. 10, pp. 2936–2946, October 2003.

[138] K. Payandehjoo, and R. Abhari, "Employing EBG structure in multi-antenna systems for improving isolation and diversity gain," IEEE Antennas and Wireless Propagation Letters, vol. 8, pp. 1162–1165, October 2009.

[139] C. C. Hsu, K. H. Lin, and H. L. Su, "Implementation of broadband isolator using metamaterial inspired resonators and a T shaped branch for MIMO applications," IEEE Transactions on Antennas and Propagation, vol. 59, no. 10, pp. 3936–3939, October 2011.

[140] D. K. Ntaikos, and T. V. Yioultsis, "Compact split ring resonator (SRR) loaded multiple input multiple output antenna with electrically small elements and reduced mutual coupling," IET Microwave Antennas and Propagation, vol. 7, Issue 6, pp. 421–429, January 2013.

[141] M. U. Khan, and M. S. Sharawi, "A 2x1 multiband MIMO antenna system consisting of miniaturized patch elements," Microwave and Optical Technology Letters, vol. 56, no. 6, pp. 1371–1375, June 2014.

[142] K. Payandehjoo, and R. Abhari, "Highly isolated uni-directional multi-slot antenna systems for enhanced MIMO performance," International Journal of RF and Microwave Computer Aided Engineering, vol. 24, no. 3, pp. 289–297, May 2014.

[143] K. Payandehjoo, and R. Abhari, "Isolation enhancement between tightly spaced compact uni-directional patch antennas on multi-layer EBG surfaces," International Journal of RF and Microwave Computer Aided Engineering, vol. 25, no. 1, pp. 30–38, January 2015.

[144] I. J. Garcia Zuazola, L. Azpilicueta, A. Sharma, H. Landaluce, F. Falcone, I. Angulo, A. Perallos, W. G. Whittow, J. M. H. Elmirghani, and J. C. Batchelor, "Band-pass filter like antenna validation in an ultra-wideband in car wireless channel," IET Communications, vol. 9, no. 4, pp. 532–540, 2015.

[145] F. Falcone, J. Illescas, E. Jarauta, A. Estevez, and J. A. Marcotegui, "Analysis of stripline configurations loaded with complementary split ring resonators," Microwave and Optical Technology Letters, vol. 55, no. 6, pp. 1250–1254, June 2013.

[146] M. Beruete, I. Campillo, J. S. Dolado, J. E. Rodrguez-Seco, E. Perea, F. Falcone, and M. Sorolla, "Very low profile and dielectric loaded feeder antenna," IEEE Antennas and Wireless Propagation Letters, vol. 6, pp. 544–548, 2007.

[147] S. Imaculate Rosaline, and S. Raghavan, "Split ring loaded broadband monopole with SAR reduction," Microwave and Optical Technology Letters, vol. 58, no. 1, pp. 158–162, January 2016.

[148] S. Raghavan, and P. Thiruvalar Selvan, "Novel compact CPW fed printed slot antenna for 5.8-GHz RFID application," Microwave and Optical Technology Letters, vol. 55, no. 12, pp. 2918–2920, December 2013.

[149] M. S. Sharawi, M. U. Khan, A. B. Numan, and D. N. Aloi, "A CSRR loaded MIMO antenna system for ISM band operation," IEEE Transactions on Antennas and Propagation, vol. 61, no. 8, pp. 4265–4274, August 2013.

[150] M. S. Sharawi, "A 5 GHz 4/8 element MIMO antenna system for IEEE 802.11 AC devices," Microwave and Optical Technology Letters, vol. 55, no. 7, pp. 1589–1594, July 2013.

[151] M. S. Sharawi, A. B. Numan, and D. N. Aloi, "Isolation improvement in a dual band element MIMO antenna system using capacitively loaded loops (CLLs)," Progress in Electromagnetics Research, vol. 134, pp. 247–266, January 2013.

[152] D. G. Yang, D. O. Kim, and C. Y. Kim, "Design of a dual band MIMO monopole antenna with high isolation using slotted CSRR for WLAN," Microwave and Optical Technology Letters, vol. 56, no. 10, pp. 2252–2257, October 2014.

[153] S. H. Hwang, T. S. Yang, J. H. Byun, and A. S. Kim, "Complementary pattern method to reduce mutual coupling in metamaterial antennas," IET Microwaves Antennas and Propagation, vol. 4, Iss. 9, pp. 1397–1405, January 2010.

[154] M. G. N. Alsath, M. Kanagasabai, and B. Balasubramanian, "Implementation of slotted meander line resonators for isolation enhancement in microstrip patch antenna arrays," IEEE Antennas and Wireless Propagation Letters, vol. 12, pp. 15–18, January 2013.

[155] H. Arun, A. K. Sharma, M. Kanagasabai, S. Velan, C. Raviteja, and M. G. N. Alsath, "Deployment of modified serpentine structure for mutual coupling reduction in MIMO antennas," IEEE Antennas and Wireless Propagation Letters, vol. 13, pp. 277–280, February 2014.

[156] Y. Cheng, Z. Sun, W. Lu, and H. Zhu "A novel compact dual-band MIMO antenna," 3rd Asia Pacific Conference on Antennas and Propagation, Harbin, China, pp. 157–160, July 2014.

[157] D. W. Browne, M. Manteghi, M. P. Fitz, and Y. R. Samii, "Experiments with compact antenna arrays for MIMO radio communications," IEEE

Transactions on Antennas and Propagation, vol. 54, no. 11, pp. 3239–3250, November 2006.

[158] S. Lung, S. Yang, K. M. Luk, H. W. Lai, A. A. Kishk, and K. F. Lee, "A dual polarized antenna with pattern diversity", IEEE Antennas and Propagation Magazine, vol. 50, no. 6, pp. 71–79, December 2008.

[159] P. Callaghan, and J. C. Batchelor, "Dual-band pin patch antenna for Wi-Fi applications," IEEE Antennas and Wireless Propagation Letters, vol. 7, pp. 757–760, August 2008.

[160] X. Ling, and R. Li, "A novel dual band MIMO antenna array with low mutual coupling for portable wireless devices," IEEE Antennas and Wireless Propagation Letters, vol. 10, pp. 1039–1042, October 2011.

[161] H. S. Singh, B. Meruva, G. K. Pandey, P. K. Bharti, and M. K. Meshram, "Low mutual coupling between MIMO antenns by using two folded shorting strips," Progress in Electromagnetics Research B, vol. 53, pp. 205–221, July 2013.

[162] B. P. Chacko, G. Augustin, and T. A. Denidni, "Uniplanar polarization diversity antenna for ultra-wideband systems," IET Microwaves, Antennas and Propagation, vol. 7, Iss. 10, pp. 851–857, March 2013.

[163] X. Wang, Z. Feng, and K. M. Luk, "Pattern and polarization diversity antenna with high isolation for portable wireless devices," IEEE Antennas and Wireless Propagation Letters, vol. 8, pp. 209–211, April 2009.

[164] J. Lu, Z. Kuai, X. Zhu, and N. Zhang, "A high isolation dual polarization microstrip patch antenna with quasi cross shaped coupling slot," IEEE Transactions on Antennas and Propagation, vol. 59, no. 7, pp. 2713–2717, July 2011.

[165] E. A. Davin, M. C. Fabres, B. B. Clemente, and F. Bataller, "Printed multimode antenna for MIMO systems," Journal of Electromagnetic Waves and Applications, vol. 25, no. 14-15, pp. 2022–2032, 2011.

[166] C. Y. Desmond Sim, "Conical beam array antenna with polarization diversity," IEEE Transactions on Antennas and Propagation, vol. 60, no. 10, pp. 4568–4572, October 2012.

[167] M. S. Sharawi, M. A. Jan, and D. N. Aloi, "Four shaped 2x2 multi standard compact multiple input multiple output antenna system for long term evolution mobile handsets," IET Microwaves Antennas and Propagation, vol. 5, Iss. 7, pp. 685–696, April 2012.

[168] W. Li, W. Lin, and G. Yang, "A compact MIMO antenna system design with low correlation from 1710 MHz to 2690 MHz," Progress in Electromagnetic Research, vol. 144, pp. 59–65, January 2014.

[169] W. Li, X. J. Chen, B. Zhang, J. H. Zhou, and B. Q. You, "Dual polarized cavity backed annular slot antenna of compact structure," Electronics Letters, vol. 50, no. 23, pp. 1655–1656, November 2014.

[170] H. S. Wong, M. T. Islam, and S. Kibria, "Design and optimization of LTE 1800 MIMO antenna," International Journal of Antennas and Propagation, research article, vol. 2014, pp. 1–10, May 2014.

[171] Y. Ding, Z. Du, K. Gong, and Z. Feng, "A four element antenna system for mobile phones," IEEE Antennas and Propagation Letters, vol. 6, pp. 655–658, November 2007.

[172] M. S. Sharawi, S. S. Iqbal, and Y. S. Faouri, "An 800 MHz 2×1 compact MIMO antenna system for LTE handsets," IEEE Transactions on Antenna and Propagation, vol. 59, no. 8, pp. 3128–3131, August 2011.

[173] J. R. Panda, and R. S. Kshetrimayum, "A printed 2.4/5.8 GHz dual-band monopole antenna with a protruding stub in the ground for WLAN and RFID applications," Progress in Electromagnetic Research, vol. 117, pp. 425–434, June 2011.

[174] J. H. Lu, and Y. H. Li, "Planar multi-band T shaped monopole antenna with a pair of mirrored L shaped strips for WLN/WiMAX applications," Progress in Electromagnetic Research C, vol. 21, pp. 33–44, April 2011.

[175] C. Yang, Y. Yao, J. Yu, P. Xie, and X. Chen, "Novel compact multiband MIMO antenna for mobile terminals," International Journal of Antennas and Propagation, research article, vol. 2012, pp. 1–9, January 2012.

[176] X. X. Xia, Q. X. Chu, and J. F. Li, "Design of a compact wideband MIMO antenna for mobile terminals," Progress in Electromagnetic Research C, vol. 41, pp. 163–174, July 2013.

[177] X. Zhao, and J. Choi, "Design of a MIMO antenna with low ECC for 4G mobile terminal," Microwave and Optical Technology Letters, vol. 56, no. 4, pp. 965–970, April 2014.

[178] Y. L. Ban, Z. X. Chen, Z. Chen, K. Kang, and J. L. Wei Li, "Decoupled closely spaced heta-band antenna array for WWAN/LTE smartphone applications," IEEE Antennas and Wireless Propagation Letters, vol. 13, pp. 31–34, January 2014.

[179] S. Shoaib, I. Shoaib, N. Shoaib, X. Chen, and C. G. Parini, "Design and performance study of a dual element multi-band printed monopole antenna array for MIMO terminals," IEEE Antennas and Wireless Propagation Letters, vol. 13, pp. 329–332, February 2014.

[180] O. Staub, J.F. Zurcher, and A. Skrivervik, "Some considerations on the correct measurement of the gain and bandwidth of electrically small

antennas," Microwave and Optical Technology Letters, vol. 17, no. 3, pp. 156–160, February 1998.

[181] C. Y. Chiu, K. M. Shum, and C. H. Chan, "A tunable via patch loaded PIFA with size reduction," IEEE Transactions on Antennas and Propagation, vol. 46, no. 1, pp. 65–71, January 2007.

[182] J. A. T. Mendez, R. A. Herrera, R. F. Leal, R. L. Miranda, and H. J. Aguilar, "IFA and PIFA size reduction by using a stub loading," International Journal of Antennas and Propagation, vol. 2013, pp. 1–7, October 2013.

[183] S. R. Best, and J. D. Morrow, "The effectiveness of space filling fractal geometry in lowering resonant frequency," IEEE Antennas and Wireless Propagation Letters, vol. 1, pp. 112–115, 2002.

[184] Shivnarayan, and B. R. Vishvakarma, "Analysis of inclined slot loaded patch for dual-band operation," Microwave and Optical Technology Letters, vol. 48, no. 12, pp. 2436–2441, December 2006.

[185] E. A. Soliman, W. De Raedt, and G. A. E. Vanderbosch, "Reconfigurable slot antenna for polarization diversity," Journal of Electromagnetic Waves and Applications, vol. 23, no. 7, pp. 905–916, 2009.

[186] H. U. Iddi, M. R. Kamarudin, T. A. Rahman, and R. Dewan, "Reconfigurable monopole antenna for WLAN/WIMAX applications," PIERS Proceedings, Taipei, pp. 1048–1051, March 2013.

[187] K. Gosalia, and G. Lazzi, "Reduced size, dual-polarized micro-strip patch antenna for wireless communications," IEEE Transactions on Antennas and Propagation, vol. 51, no. 9, pp. 2182–2188, September 2003.

[188] G. P. Maddani, N. M. Sameena, and S. N. Mulgi, "Wide bandwidth, high gain slot loaded rectangular microstrip antennas," Microwave and Optical Technology Letters, vol. 52, no. 12, pp. 2841–2845, December 2010.

[189] Z. W. Yu, G. M. Wang, X. J. Gao, and K. Lu, "A novel small size single patch micro-strip antenna based on koch and sierpinski fractal shapes," Progress in Electromagnetics Research Letters, vol. 17, pp. 95–103, 2010.

[190] C. Arora, S. S. Pattnaik, and R. N. Baral, "SRR inspired microstrip patch antenna array," Progress in Electromagnetics Research C, vol. 58, pp. 89–96, 2015.

[191] Y. Xie, L. Li, C. Zhu, and C. Liang, "A novel dual-band patch antenna with complementary split ring resonators embedded in the ground plane," Progress in Electromagnetics Research Letters, vol. 25, pp. 117–126, 2011.

[192] P. L. Shu, and Q. Y. Feng, "Design of a compact quadband hybrid antenna for compass/WiMAX/WLAN applications," Progress in Electromagnetics Research, vol. 138, pp. 585–598, April 2013.

[193] N. Herscovici, "New considerations in the design of microstrip antennas," IEEE Transactions on Antennas and Propagation, vol. 46, no. 6, pp. 807–812, June 1998.

[194] Leeladhar Malviya, Rajib K. Panigrahi, and M. V. Kartikeyan, "MIMO antennas with diversity and mutual coupling reduction techniques: a review," International Journal of Microwave and Wireless Technologies (IJMWT), vol. 9, pp. 1763–1780, May 2017.

[195] K. Sankar, R. Bargavi, and Arivumani S. Samson, "Single layer dual-band G-shaped patch antenna," International Conference on Communication and Signal Processing, pp. 636–639, India, April 2014.

[196] L. D. Malviya, J. Malik, R. K. Panigrahi, and M. V. Kartikeyan, "Design of a compact MIMO antenna with polarization diversity technique for wireless communication," International Conference on Microwave, Optical and Communication Engineering (ICMOCE), Bhubaneswar, India, pp. 21–24, December 2015.

[197] D. C. Chang, B. H. Zeng, and J. C. Liu, "Reconfigurable angular diversity antenna with quad corner reflector arrays for 2.4 GHz applications," IET Microwaves Antennas and Propagation, vol. 3, Iss. 37, pp. 522–528, April 2009.

[198] Leeladhar Malviya, Rajib K. Panigrahi, and Machavaram V. Kartikeyan, "A 2×2 Dual-band MIMO antenna with polarization diversity for wireless applications," Progress in Electromagnetics Research C, vol. 61, pp. 91–103, January 2016.

[199] A. T. Hassan, and M. S. Sharawi, "Four element half circle shape printed MIMO antenna," Microwave and Optical Technology Letters, vol. 58, no. 12, pp. 2990–2992, September 2016.

[200] J. Y. Lee, S. H. Kim, and J. H. Jang, "Reduction of mutual coupling in planar multiple antenna by using 1-D EBG and SRR structures," IEEE Transactions on Antennas and Propagation, vol. 63, no. 9, pp. 4194–4198, September 2015.

[201] A. Ramachandran, S. V. Pushpakaran, M. Pezholil, and V. Kesavath, "A four port MIMO antenna using concentric square ring patches loaded with CSRR for high isolation," IEEE Antennas and Wireless Propagation Letters, vol. 15, pp. 1196–1199, April 2016.

[202] X. Wang, W. Chen, J. Zheng, Y. Shen, and Z. Yeng, "Design and preliminary evaluation of a compact four element terminal multiple input multiple output antenna for receiving antenna selection," IET Microwaves Antennas and Propagation, vol. 5, Issue 7, pp. 756–763, May 2011.

[203] J. Malik, A. Patnaik, and M. V. Kartikeyan, "Novel printed MIMO antenna with pattern and polarization diversity," IEEE Antennas and Wireless Propagation Letters, vol. 14, pp. 739–742, March 2015.

[204] T. Huang, Y. Yu, and L. Yi, "Design of highly isolated compact antenna array for MIMO applications," International Journal of Antennas and Propagation, vol. 2014, pp. 1–5, November 2014.

[205] Z. Z. Abidin, Y. Ma, R. A. Abd Alhameed, K. N. Ramli, D. Zhou, M. S. Bin Melha, J. M. Noras, and R. Halliwell, "Design of 2 × 2 U shape MIMO slot antennas with EBG material for mobile handset applications," Progress in Electromagnetic Research Symposium, vol. 7, no. 1, pp. 81–84, March 2011.

[206] W. Cao, A. Liu, B. Zhang, T. Yu, and Z. Quan, "Dual-band spiral patch slot antenna with omnidirectional CP properties," IEEE Transactions on Antennas and Propagation, vol. 61, no. 4, pp. 2286–2289, April 2013.

[207] L. D. Malviya, R. K. Panigrahi, and M. V. Kartikeyan, "Pattern diversity based MIMO antenna for low mutual coupling," IEEE Applied Electro-magnetic Conference (AEMC), Guwahati, India, pp. 96–97, December 2015.

[208] Leeladhar Malviya, R. K. Panigrahi, and M. V. Kartikeyan, "A multistandard, wide-band 2 × 2 compact MIMO antenna with ground modification techniques," International Journal of Microwave and Optical Technology (IJMOT), vol. 11, no. 4, pp. 259–267, July 2016.

[209] J. Wang, Z. Lv, and X. Li, "Analysis of MIMO diversity improvement using circular polarized antenna," International Journal of Antennas and Propagation, vol. 2014, pp. 1–9, January 2014.

[210] P. N. Rao, and N. V. S. N. Sarma, "Fractal boundary circularly polarized single feed microstrip antenna," Electronics Letters, vol. 44, no. 12, pp. 713–714, June 2008.

[211] V. V. Reddy, and N. V. S. N. Sarma, "Compact circularly polarized asymmetrical fractal boundary microstrip antenna for wireless applications," IEEE Antennas and Wireless Propagation Letters, vol. 13, pp. 118–121, 2014.

[212] V. V. Reddy, and N. V. S. N. Sarma, "Tri-band circularly polarized koch fractal boundary microstrip antenna," IEEE Antennas and Wireless Propagation Letters, vol. 13, pp. 1057–1060, 2014.

[213] S. R. Rama, D. Vakula, and N. V. S. N. Sarma, "Multi-band multipolarized planar antenna for WLAN/WiMAX applications," 31st International Review of Progress in Applied Computational Electromagnetics (ACES), pp. 1–2, March 2015.

[214] A. K. Gautam, and B. K. Kanaujia, "A novel dual-band asymetric slit with defected ground structure microstrip antenna for circular polarization operation," Microwave and Optical Technology Letters, vol. 55, no. 6, pp. 1198–1201, February 2013.

[215] S. kumar, B. K. Kanaujia, A. Sharma, M. Khandelwal, and A. K. Gautam, "Single feed cross slot loaded compact circularly polarized microstrip antenna for indoor WLAN applications," Microwave and Optical Technology Letters, vol. 56, no. 6, pp. 1313–1317, June 2014.

[216] S. A. Rezaeieh, S. Simsek, and J. Pourahmadazar, "Design of a compact broadband circularly polarized slot antenna for wireless applications," Microwave and Optical Technology Letters, vol. 55, no. 2, pp. 413–418, February 2013.

[217] J. H. Lu, and Y. H. Liu, "Novel dual-band design of planar slot array antenna for 4G LTE/WiMAX access points," Microwave and Optical Technology Letters, vol. 54, no. 5, pp. 1193–1196, May 2012.

[218] Y. Jiang, W. Geyi, and H. Sun, "A new focused antenna array with circular polarization," Microwave and Optical Technology Letters, vol. 57, no. 12, pp. 2936–2939, December 2015.

[219] S. Karamzadeh, B. S. Virdee, V. Rafii, and M. Kartal "Circularly polarized slot antenna array with sequentially rotated feed network for broadband application," International Journal of RF and Microwave Computer Aided Engineering , vol. 25, no. 4, pp. 358–363, May 2015.

[220] M. K. A. Nayan, M. F. Jamlos, and M. A. Jamlos, "MIMO circular polarization array antenna with dual coupled 90° phased shift for point to point applications," Microwave and Optical Technology Letters, vol. 57, no. 4, pp. 809–814, April 2015.

[221] M. K. A. Nayan, M. F. Jamlos, H. Lago, and M. A. Jamlos, "Two port circular polarized antenna array for point to point applications," Microwave and Optical Technology Letters, vol. 57, no. 10, pp. 2328–2332, October 2015.

[222] L. J. Gang, G. X. Jun, and Z. H. Zhou, "The design of broadband circularly polarized planar microstrip antenna array," Microwave and Optical Technology Letters, vol. 49, no. 12, pp. 2936–2939, December 2007.

[223] M. K. A. Nayan, M. F. Jamlos, and M. A. Jamlos, "Circularly polarized MIMO antenna array for point to point communication," Microwave and Optical Technology Letters, vol. 57, no. 1, pp. 242–247, January 2015.

[224] J. Abraham, T. Mathew, and C. K. Aanandan, "A novel proximity fed gap coupled microstrip patch array for wireless applications," Progress in Electromagnetic Research C, vol. 61, pp. 171–178, January 2016.

[225] B. A. Khawaja, M. A. Tarar, T. Tauqeer, F. Amir, and M. Mustaqim, "A 1×2 triple band printed antenna array for use in next generation flying Ad-hoc networks," Microwave and Optical Technology Letters, vol. 58, no. 3, pp. 606–610, March 2016.

[226] C. Sun, H. Zheng, and Y. Liu, "Analysis and design of a low cost dual-band compact circularly polarized antenna for GPS application," IEEE Transactions on Antennas and Propagation, vol. 64, no. 1, pp. 365–370, January 2016.

[227] A. H. Haghparast, and G. Dadashzadeh, "A dual-band polygon shaped CPW-fed planar monopole antenna with circular polarization and isolation enhancement for MIMO aplications," 9th European Conference on Antennas and Propagation (EuCAP), pp. 1–4, May 2015.

[228] Leeladhar Malviya, Rajib K. Panigrahi, and Machavaram V. Kartikeyan, "Circularly polarized 2×2 MIMO antenna for WLAN applications," Progress in Electromagnetics Research C, vol. 66, pp. 97–107, July 2016.

[229] C. B. Dietrich, K. Dietze, J. R. Nealy, and W. L. Stutzman, "Spatial, polarization, and pattern diversity for wireless hand-held terminals," IEEE Transactions on Antennas and Propagation, vol. 49, no. 9, pp. 1271–1281, September 2001.

[230] A. K. Skrivervik, J. F. Zurcher, O. Staub, and J. R. Mosig, "PCS antenna design: The challenge of miniaturization," IEEE Antennas and Propagation Magazine, vol. 43, no. 4, pp. 12–27, August 2001.

[231] A. Mishra, P. Singh, N. P. Yadav, J. A. Ansari, and B. R. Vishvakarma, "Compact shorted micro-strip patch antenna for dual band operation," Progress in Electromagnetic Research C, vol. 9, pp. 171–182, 2009.

[232] A. Foudazi, A. Mallahzadeh, and S. M. A. Nezhad, "A triple band WLAN/WiMAX printed monopole antenna for MIMO applications," Microwave and Optical Technology Letters, vol. 54, pp. 1321–1325, May 2012.

[233] N. I. M. Elamin, T. A. Rahman, and A. Y. Abdulrahman, "New adjustable slot meander patch antenna for 4G handheld devices," IEEE Antennas and Wireless Propagation Letters, vol. 12, pp. 1077–1080, September 2013.

[234] J. F. Li, and Q. X. Chu, "A compact dual-band MIMO antenna of mobile phone," Journal of Electromagnetic Waves and Application, vol. 25, pp. 1577–1586, 2011.

[235] W. S. Chen, and B. Y. Lee, "Novel printed monopole antenna for PDA phone and WLAN applications," Journal of Electromagnetic Waves and Application, vol. 23, pp. 2073–2088, 2009.

[236] Y. K. Choukiker, S. K. Sharma, and S. K. Behera, "Hybrid fractal shape planar monopole antenna covering multiband wireless communications with MIMO implementation for handheld mobile devices," IEEE Transactions on Antennas and Propagation, vol. 62, no. 3, pp. 1483–1488, March 2014.

[237] Y. Torabi, A. Bahri, and A. R. Sharifi, "A novel metamaterial MIMO antenna with improved isolation and compact size based on LSRR resonator," IETE Journal of Research, vol. 62, Iss. 1, pp. 106–112, 2016.

[238] R. Karimian, H. Oraizi, S. Fakhte, and M. Farahani, "Novel F shaped quad-band printed slot antenna for WLAN and WiMAX MIMO systems," IEEE Antennas and Wireless Propagation Letters, vol. 12, pp. 405–408, March 2013.

[239] M. Sonkki, D. Pfeil, V. Hovinen, and K. R. Dandekar, "Wide-band planar four element linear antenna array," IEEE Antennas and Wireless Propagation Letters, vol. 13, pp. 1663–1666, August 2014.

[240] Leeladhar Malviya, Rajib K. Panigrahi, and M. V. Kartikeyan, "Four element planar MIMO antenna design for long term evolution operation," IETE Journal of Research (Taylor and Francis), vol. 64, no. 3, pp. 367–373, August 2017.

[241] A. Alexiou and M. Haardt, "Smart antenna technologies for future wireless systems: trends and challenges," IEEE Communications Magazine, vol. 42, no. 9, pp. 90–97, September 2004.

[242] H. Wei, D. Wang, H. Zhu, J. Wang, S. Sun, and X. You, "Mutual coupling calibration for multiuser massive MIMO systems," IEEE Transactions on Wireless Communications, vol. 15, no. 1, pp. 606–619, January 2016.

[243] S. A. Rezaeieh, and N. Pouyanfar, "Double channel triple band MIMO antenna with high isolation performance and pattern diversity for wireless applications," Microwave and Optical Technology Letters, vol. 54, no. 12, pp. 2689–2691, December 2012.

[244] Y. Wang and Z. Du, "A printed dual-antenna system operating in the GSM1800/GSM1900/UMTS/LTE2300/LTE2500/2.4 GHz WLAN bands for mobile terminals," IEEE Antennas and Wireless Propagation Letters, vol. 13, pp. 233–236, February 2014.

[245] S. Shoaib, I. Shoaib, N. Shoaib, X. Chen, and C. G. Parini, "MIMO antennas for mobile handsets," IEEE Antennas and Wireless Propagation Letters, vol. 14, pp. 799–802, March 2015.

[246] Y. G. Kim, and S. O. Park, "A wideband dual polarization base station antenna," Microwave and Optical Technology Letters, vol. 57, no. 1, pp. 22–26, January 2015.

[247] Leeladhar Malviya, Rajib K. Panigrahi, and M. V. Kartikeyan, "A low profile planar MIMO antenna with polarization diversity for LTE 1800/1900 applications," Microwave and Optical Technology Letters (MOTL), vol. 59, no. 3, pp. 533–538, March 2017.

[248] Leeladhar Malviya, Rajib K. Panigrahi, and M. V. Kartikeyan, "Proximity coupled MIMO antenna for WLAN/WiMAX applications," IEEE Asia Pacific Microwave Conference 2016 (APMC), Delhi, India, pp. 1–4, December 2016.

[249] Gourab Das, Anand Sharma, and Ravi Kumar Gangwar, "Wideband self-complementary hybrid ring dielectric resonator antenna for MIMO applications," IET Microwave Antennas and Propagation, vol. 12, no. 1, pp. 108–114, January 2018.

[250] Sumeet S. Bhatia, Aditi Sahni, and Shashi B. Rana, "A novel design of compact monopole antenna with defected ground plane for wideband applications," Progress in Electromagnetics Research M, vol. 70, pp. 21–31, June 2018.

[251] Saber Soltani, Parisa Lotfi, and Ross D. Murch, "A dual-band multiport MIMO slot antenna for WLAN applications," IEEE Antennas and Wireless Propagation Letters, vol. 16, pp. 529–532, July 2016.

[252] Rizwan Masood, Christian Person, and Ronan Sauleau, "A dual-mode, dual port pattern diversity antenna for 2.45 GHz WBAN," IEEE Antennas and Wireless Propagation Letters, vol. 16, pp. 1064–1067, October 2016.

[253] Yashika Sharma, Debdeep Sarkar, Kushmanda Saurav, and Kumar Vaibhav Shrivastava, "Three element MIMO antenna system with pattern and polarization diversity for WLAN applications," IEEE Antennas and Wireless Propagation Letters, vol. 16, pp. 1163–1166, November 2016.

[254] Ye-Hai Bi, Yihong Qi, Jun Fan, and Wei Yu, "A planar low profile antenna (PLMA) design for wireless terminal achieving between intra-system EMC and isolation in multi-antenna system," IEEE Transactions on Electromagnetic Compatibility, vol. 59, pp. 980–987, October 2017.

[255] Leeladhar Malviya, Rajib K. Panigrahi, and M. V. Kartikeyan, "2×2 MIMO antenna for ISM band application," 11th International Conference on Industrial and Information Systems (ICIIS 2016), Roorkee, India, pp. 794–797, December 2016.

[256] Leeladhar Malviya, M. V. Kartikeyan, and Rajib K. Panigrahi, "Offset planar MIMO antenna for omnidirectional radiation patterns," International Journal of RF and Microwave Computer Aided Engineering, pp. 1–9, October 2018.

[257] Leeladhar Malviya, M. V. Kartikeyan, and Rajib K. Panigrahi, "Multi-standard, multi-band planar MIMO antenna with diversity effects for wireless applications," International Journal of RF and Microwave Computer Aided Engineering, pp. 1–8, September 2018.

Index